The Australian Native Bee Book

Keeping stingless bee hives for pets, pollination and sugarbag honey

Winner of a 2016 Whitley Award for
Australasian Zoological Literature

by Tim Heard

Sugarbag Bees

PUBLISHED BY Sugarbag Bees, 2016

© SUGARBAG BEES
473 Montague Road, West End, Queensland 4101 Australia

FOR ORDERS: contact tim@sugarbag.net
or order online at nativebeebook.com.au

PHOTOGRAPHY COPYRIGHT © All photographs, illustrations, diagrams and drawings are copyright of the owner as stated in the caption, otherwise are copyright of Tim Heard.

DESIGNED BY Glenbo Craig

TEXT COPYRIGHT © Tim Heard 2016

PRINTED BY Express Print and Mail, Maroochydore, Australia

National Library of Australia Cataloguing-in-Publication entry:

Heard, Tim A., 1958 – author.

The Australian Native Bee Book: Keeping stingless bee hives for pets, pollination and sugarbag honey / Tim Heard

ISBN 978-0-646-93997-1 (paperback)

Includes index

Bee culture, Australian native stingless bees, Pollination, Honey.

COVER PHOTOS

Front cover Brood comb of *Tetragonula carbonaria*.
IMAGE **MIKAYLA LAMBERT**

Front flap *Tetragonula* stingless bee visiting a blueberry flower.

Back cover Top – A stingless bee (*Tetragonula* sp.) foraging for pollen on wonga wonga vine flowers.
IMAGE **LAURENCE SANDERS**

2nd top – A painted hive at Honeycomb Valley, NSW.

3rd top – Stingless bee hives on a macadamia farm.
IMAGE **GIORGIO VENTURIERI**

For Katina

Male and female peacock carpenter bees on leptospermum flowers.
IMAGE CORINNE JORDAN

Acknowledgements

I wish to sincerely thank many people for their help in producing this book.

To Australia's keepers of stingless bees, I am deeply grateful to you for teaching me so much. In particular, I wish to acknowledge Peter Davenport, Tony Goodrich, John Klumpp, Russell and Janine Zabel, Rob Raabe, Frank Adcock, Denis Shepherd, Wendy Forno, Tom Carter, Chris Fuller, Mark Grosskopf, Allan Beil, Cec Heather, Steve Maginnity and Peter Clarke.

Dr Anne Dollin (Australian Native Bee Research Centre) provided valuable comments on the manuscript and co-authored the chapter "The stingless bees of Australia". Anne is a fine colleague and good friend. I have learnt so much from her as our paths have crossed and we have travelled together on numerous shared projects over several decades. I cannot overstate the fundamental contribution of Anne, and husband Les Dollin, to our knowledge of native bees in Australia.

Thank you also to the many photographers who have allowed me to use their images: Nadine Andersen, Dr Denis Anderson, Jan Anderson, Damien Andrews, Melissa Ballantyne, Justin Bartlett, Gail Bruce, Carole Bryant, Eva Cachin, Tom Carter, Dianne Clarke, Jerry Coleby-Williams, Dan Coughlan, Glenbo Craig, Dr Bronwen Cribb, Dr Paul Cunningham, Dr Anne Dollin, James Dorey, Dr Michael S. Engel, Chris Fuller, Caroline Gardam, Dr Ros Gloag, Dr Julia Groening, Dr Megan Halcroft, Katina Heard, Dr Katja Hogendoorn, Jean and Fred Hort, Bernhard Jacobi, Corinne Jordan, Benjamin Kaluza, Daniel Klaer, John Klumpp, Mikayla Lambert, Dr Danielle LeLagadec, Dr Remko Leijs, Dr Robert Matthews, Dr C. Flavia Massaro, Dr David Merritt, Marc Newman, Amanda Norton, Claudia Rasche, Malcolm Ricketts, Reiner Richter, Professor Simon Robson, Linda Rogan, Laurence Sanders, Jenny Shanks, Dr Aung Si, Erica Siegel, Dr Tobias Smith, Don Smith, Marcio Sztutman, Richard Tanner, Jenny Thynne, Eric Tourneret, Dr Myfany Turpin, Dr Giorgio Venturieri, Dr Ken Walker, Professor Helen Wallace, Jeff Willmer, Dr Alan Yen, Russell Zabel. The credit for each image is provided in its caption; where no credit is given, the photo is mine.

Gina Cranson sketched and coloured the magnificent artwork that not only looks gorgeous but serves to illustrate in a way that photographs cannot. Daniel Klaer created the diagrams of hive box design. Dr Tobias Smith helped draw the global maps and prepared the Index.

Many people have lent a generous hand by editing this entire book or specific chapters. I wish to express my deep gratitude to Dr Anne Dollin (Australian Native Bee Research Centre), Dr Giorgio Venturieri (EMBRAPA Brazil), Dr Tobias Smith (University of Queensland), Dr Ros Gloag (University of Sydney), Dr Danielle Lloyd-Prichard (University of Newcastle), Jeff Willmer and John Klumpp for their general reviewing. Dr Michael Batley (Australian Museum) provided deep insights into the first two chapters. Professor Ben Oldroyd (University of Sydney) reviewed the chapter on social behaviour. Professor Madeleine Beekman (University of Sydney) revised the nesting chapter. Dr Claus Rasmussen (Aarhus University, Denmark) checked the section on biogeography. Professor Helen Wallace (University of the Sunshine Coast) reviewed the section on cadaghi. Dr Elizabeth Frost (E. Frost Apicultural Services) checked the information on honey bees. Dr Ken Walker (Museum Victoria) revised the chapter on bee diversity. Professor James Nieh (University of California San Diego) checked the communication section. Dr Myfany Turpin (University of Sydney) reviewed the section on Indigenous peoples and stingless bees. Dr Denis Anderson (Bees Downunder) gave the section on bee diseases a full medical. Dr Chris Burwell (Queensland Museum) scrutinised the chapter on natural enemies. Dr Saul Cunningham (CSIRO) cast his astute eye over the bees and pollination chapter. Dr Megan Halcroft (Bees Business) reviewed everything related to her beloved *Austroplebeia*. Jenny Shanks (University of Western Sydney) provided the details on the stingless bee brood diseases. Dr Katja Hogendoorn (University of Adelaide) checked management of solitary bees. Dr C. Flavia Massaro (University of the Sunshine Coast) analysed the sticky business of resin and wax. The following beekeepers also gave me their valuable comments: Russell Zabel, Dean Haley, Peter Clarke, Tony Goodrich, Chris Fuller and Sarah Hamilton. Corrie Macdonald edited the final drafts and greatly improved structure, grammar and readability.

Claudia Rasche ran the Sugarbag enterprise for 12 months, freeing me up to write this book.

I give special thanks to Glenbo Craig for his ideas, guidance, publishing advice, design and photography. Glenbo has to be the most inspiring and encouraging person I have ever worked with!

Foreword

By Tim Low, biologist and prize-winning author of seven books, including *Where Song Began*.

Ask Australians to list examples of their native wildlife, and most people won't mention bees, even though Australia has around 2,000 species. These don't attract the attention they should because, as small invertebrates, they aren't as noticeable as, say, parrots and possums. And, as visitors to flowers, they are less obvious in our gardens and forests than introduced honey bees. The well publicised global decline of honey bees draws even more attention away from native bees.

Until now, another problem has been that information about our bees hasn't been readily available. This new book changes that dramatically. Australia has any number of guides to its vertebrates, but coverage of insects – apart from butterflies – has been patchy. In writing this book, Tim Heard has provided an important service, by showing how interesting our native bees are and how amenable they are to being supported in the garden.

Many years ago Tim installed stingless bee hives around my house. The bees have thrived on eucalypt blossom and other local foods. They are my insect pets. I would be going too far if I said that as pets they were exciting or affectionate, but there is something relaxing and reassuring about their endless gentle journeys to and from garden flowers. To have a hive on the verandah, and another beside the front door, is an interesting way to get closer to nature. And, as Tim explains, their honey is 'unique and exquisite'. I offer a taste to any visitor curious about the hives.

Many Australians entertain romantic notions about European honey bees, admiring them as little helpers that toil away to provide honey and pollinate our crops. This romance hinders awareness of the environmental harms they cause. Native stingless bees are more deserving of our affections – and not only because they don't sting. Tim's passion for them shows in this engaging book, which will go a long way towards persuading Australians to acknowledge bees as part of our wildlife, and to think native when they think of bees.

Contents

IMAGE **DIANNE CLARKE**

vi

Keeping native stingless bees is a hot topic in Australia for commercial, environmental and recreational reasons. You can do something about the decline of pollinators by conserving native bees.

Tetragonula carbonaria forager landing at the nest entrance with a load of resin in her pollen baskets.
IMAGE **JAMES DOREY**

Preface

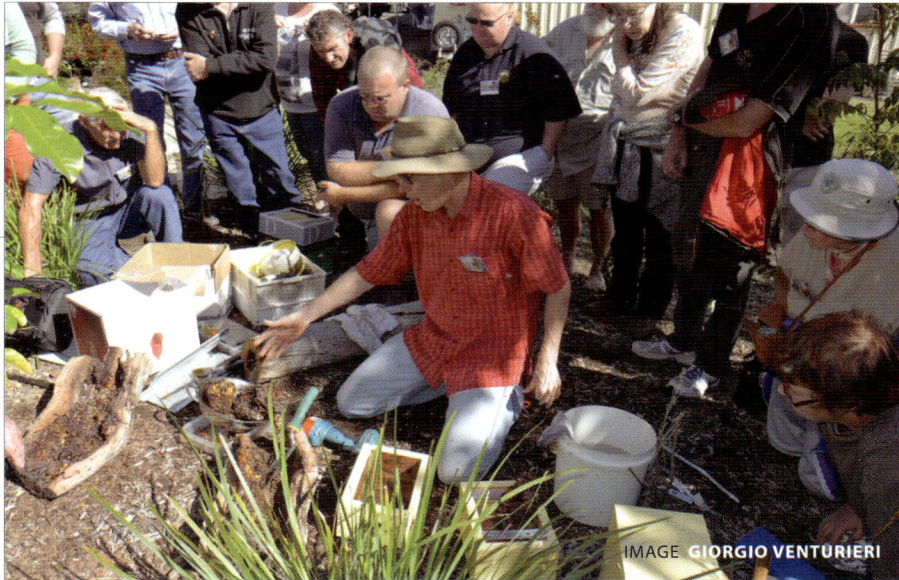

IMAGE **GIORGIO VENTURIERI**

In this book, you will find the resources on Australian native bees that I have been presenting in workshops in Australia for the last 20 years, but extended beyond what is possible in a one-day workshop. The workshop starts with the general biology of bees and finishes with practical details on the keeping of native stingless bees. This book follows that structure. I focus on Australia's social stingless bees because these are the ones of greatest interest to people, at least in the warmer areas where they naturally occur. These captivating and useful insects are the subject of a new wave of beekeeping. Thousands of people are new converts to keeping these bees as pets, for small-scale home honey production, and for pollination. They are not only great pollinators for the home garden but increasingly for commercial-scale pollination of crops.

Part 1 (the first seven chapters) of this book is devoted to building the foundations of bee biology.

Beekeeping is applied insect science, and the deeper the understanding, the better the result.

The enjoyment of beekeeping is also enhanced by a greater knowledge of the workings of bees and an appreciation of their role in the natural world.

However, if your primary objective is a practical one and you wish to skip straight to Part 2 (seven chapters on beekeeping), you should still be able to follow the information there.

The book finishes with pollination in Part 3 (the final two chapters). I also explore related topics in feature sections, including Indigenous uses of stingless bees, propolis, recipes using sugarbag honey and some of the amazing stingless bee initiatives going on around the country.

In the course of explaining bee biology, I give a brief introduction to solitary and semi-social native bees. I also cover introduced honey bees, especially in chapters 3 and 4.

Although our native stingless bees produce honey, the term "honey bees" is used exclusively here for the European and Asian species of *Apis*. European honey bees are now well entrenched in most Australian natural ecosystems, are extremely economically significant, are well studied and understood, are familiar to many people, and provide a great basis to compare and contrast with our native stingless bees.

I invite you to enter the intriguing world of bees.

Tim

Here's where you get to meet the bees in all their glorious diversity!

You will learn all about bees' wonderful, weird and downright fascinating lives. You will read about the social life of bees, how they build their complex nests, and how they rear their young – all of which will set you up nicely for keeping them in boxes.

You will also discover how bees forage – useful information if you want to use bees for pollination. You'll find out where bees came from in evolutionary terms, and how our island continent ended up sharing species with countries on the other side of the world. Plus what makes Australia's stingless bees unique.

This first part also details all the species of stingless bees that we have in Australia, where they occur and how to identify them; this leads into Part 2, where we investigate how to keep stingless bees.

BELOW Male blue-banded bee. IMAGE **ERICA SIEGEL**

RIGHT *Tetragonula carbonaria* foraging worker returning to nest with pollen load.
IMAGE **JAMES DOREY**

Part ONE

Understanding bees

1 Bee basics

IMAGE
REMKO LEIJS

Bees evolved from wasps when one species of wasp began to feed its young plant pollen instead of other insects.

Over time, the first bee species evolved into many species and forged a major partnership with the flowering plants.

Most bees are solitary but a few species have evolved very complex social behaviour.

Bees feed on protein-rich pollen supplemented by energy-rich nectar, also from flowers.

Adult female bees have an array of body structures to help them carry pollen back to the nest.

Young bees, called larvae, are grubs that live concealed in cells in the nest.

1.1 What are bees?

Bees belong to the **Hymenoptera**, a group of insects that also contains the ants and wasps. Bees are most closely related to wasps. **Wasps** are predators and parasites on other animals, mainly insects. The ancestral bee appeared about 120 million years ago, when a species of carnivorous wasp switched its diet from meat to plant pollen. So bees are really vegetarian wasps! The scientific name for bees is **Anthophila**, which means "flower lovers".

This new way of life proved to be so successful that, over millions of years, bees evolved into a hugely diverse group occupying most parts of the planet. Their timing was ideal, as flowering plants had recently emerged. Thus the foundations were laid for one of nature's great partnerships: flowering plants provided food for bees, and, in return, bees moved pollen (the male sex cells of plants) from one plant to another (a process called "pollination").

Most wasps and bees are solitary insects. Solitary wasps and bees nest in similar ways (Figure 1-1, Figure 1-2). An adult female, working alone, finds a site and constructs the nest. She mates with a male to fertilise her eggs. But males do not contribute to raising young at all; the females are the ultimate single mothers. The nest consists of one or many cells. The nests are usually in a protected position such as in soil or in hollow plant stems.

The female collects food to provision each cell. Female wasps provide prey for their young, but bees combine pollen and honey to form the larval food, either as a solid mass or a semi-fluid paste. The female then deposits an egg on the provisions and closes the cell. She repeats that process as many times as she can before she dies. Her egg hatches and the resulting **larva** consumes the food, grows, pupates and emerges as an adult.

FIGURE 1-1 A solitary wasp and its nest. This wasp has built a mud nest, stocked the nest with caterpillars, and laid its egg in the nest. The image on the right shows the nest detached and turned over to show its contents. The ancestral bee may have looked like this wasp.
IMAGE **ROBERT MATTHEWS**

FIGURE 1-2 Life cycle and nest of the solitary bee *Lipotriches australica*.
A An adult bee.
B Nest consisting of a long tunnel and side cells in soil (exposed and made visible with talcum powder).
C Cell provisioned with food and egg laid on top.
D Part-grown larva and its remaining food.
E Pale young pupa and dark mature pupa in their cells.
IMAGES **REMKO LEIJS**

EGG on brood food

LARVA with food

PUPAE

Bee basics • 3

1.2 What bees need

Bees get all the food they need from flowers. Both larvae and adult bees eat pollen and nectar. **Pollen** is especially important for larval bees, which need its high protein content to grow their bodies from egg to adult. The sugary **nectar** is vital for adult bees, which have high energy requirements. Bees may also collect **nesting materials** such as resin, leaves or mud from their environment.

Nectar is secreted by plants in organs called **nectaries,** normally found at the base of flowers (Figure 1-3). The nectar is presented in various ways; sometimes it is located deep in tubes, or it may be in open dishes. Bees suck up the sweet substance and either transport it back to the nest or digest it for an instant energy hit. The sugar content of nectar varies a lot, anywhere from 10% to 70%. The optimum sugar concentration for the maximum rate of energy intake by bees from flowers is 60%. Above 60%, the nectar becomes very viscous, which makes it slow and difficult to ingest.

In addition to sugar, nectar may contain amino acids, proteins, lipids, enzymes, organic acids, and antioxidants such as vitamin C. Nectar may even contain drugs, such as caffeine, that can manipulate the bee's behaviour. Taste receptors on bees' tongues allow them to assess the quality of nectar.

Pollen is also collected from flowers, where it is produced in the **anthers**, the male reproductive organs of plants. Bees use a variety of behaviours to get themselves thoroughly dusted with pollen by making contact with the anthers in various ways (Figure 1-3). The hairs on their bodies help to trap the pollen. Bees even carry a weak positive **electrostatic charge,** which is the opposite of the negative charge of most pollen, resulting in pollen being attracted to their bodies.

Pollen normally contains about 20–35% protein, depending on the plant species. Pollen may also contain starch, lipids, vitamins and minerals. Pollen grains of different plant species vary in size, shape and in the patterns on their surface. This allows researchers to identify pollen grains on bees, or stored in nests, or even from bee faeces, to know which plant species the bees are visiting.

Pollen from a diversity of sources is optimal for the bee diet. Pollen from different species of plants contains different types and amounts of amino acids, the building blocks of protein. A combination of different pollens creates a more balanced bee diet, producing bees that are healthier, live longer, work harder and are maybe even smarter. One of the reasons that honey bees are struggling in North America and Europe is that they are kept in human-modified areas that are monocultures of one or few plant species.

To obtain pollen and nutritional diversity, most species of bees are **generalists** that forage opportunistically on the flowers of many plant species that are available when and where they are active. The highly social bees are particularly broad in their tastes, as they have long-lived colonies active most of the year. But some bee species are **specialists**, adapted to one or a few plant species. (Read more about how bees exploit their environment to meet their vital needs in Chapter 5 "Foraging behaviour of stingless bees".)

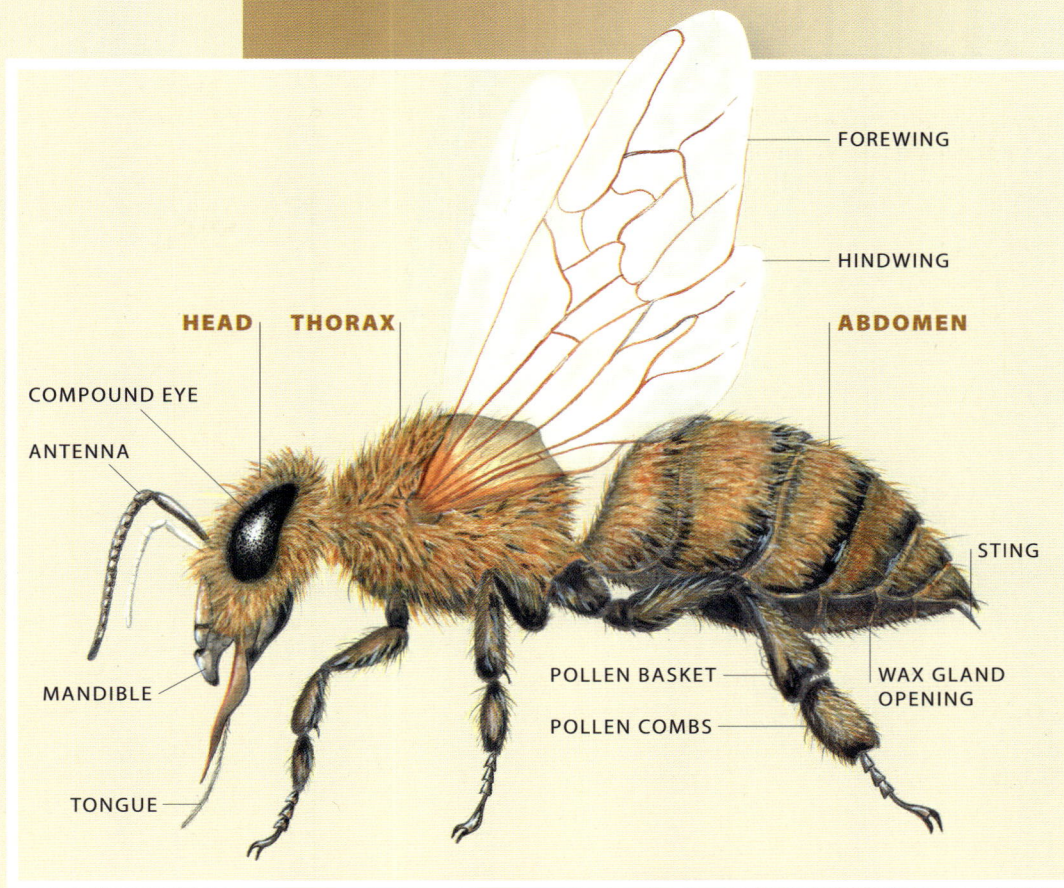

FOREWING

HINDWING

HEAD **THORAX** **ABDOMEN**

COMPOUND EYE

ANTENNA

STING

MANDIBLE

POLLEN BASKET

WAX GLAND OPENING

POLLEN COMBS

TONGUE

1.3 The ABC of bee anatomy

FIGURE 1-4
The external parts of the bee. We need to know about these to really understand how bees work.
ARTWORK **GINA CRANSON**

FIGURE 1-3
Flowers provide bees with both pollen and nectar. The flower on the left is dripping sweet nectar, while that on the right is producing abundant pollen.
IMAGES
GIORGIO VENTURIERI

Bees are superbly adapted for their way of life, and this is evident in their body structure (Figure 1-4). Bees, like all insects, have an **exoskeleton**, a hard external shell that supports and protects the body. The body is divided into **head**, **thorax** and **abdomen**.

The **head** is the nerve centre and houses an impressive array of sensory organs. The large **compound eyes** couple with the large optic lobe of the brain to provide excellent vision. Bees use that vision to locate resources and find their way home. Their **antennae** are wired with organs of touch, smell and taste. These extremely useful senses have a myriad of functions. For example, foraging bees use smells to find resources. The sense of touch serves bees well in the darkness of the nest, allowing them to find their way around and to build complex nest structures. Taste is used by guard bees at the nest entrance so that any other insect attempting to enter will be detected, perceived to be an imposter, and repelled.

The head bears a pair of strong **mandibles** or **jaws**. These can cut, dig, chew and grip, and so are useful for defending, building nests, and foraging. The **tongue** of the long-tongued bees is in the form of a hollow tube, their built-in drinking straw. Nectar is sucked up from the otherwise inaccessible depths of flowers by this tidy little tool. The short-tongued bee's tongue is adapted for lapping, not sucking. In both cases, the nectar is drawn into the crop, to be transported back to the nest.

Bee basics • 5

SCOPA LOADED
WITH POLLEN

FIGURE 1-5

The scopa of female leaf-cutter bees (*Megachile punctata*, Megachilidae) is a large flat area covered in long, pollen-carrying hairs, under the abdomen.
IMAGE **ERICA SIEGEL**

Bees evolved a broad diversity of morphological changes to better deal with the change of diet from prey to pollen.

In particular, most species sport a structure in which the pollen is packed for secure transport back to the nest. It can be either a patch of special hairs (a **scopa**) or a pollen basket (a **corbicula**). The scopa is usually under the abdomen (Figure 1-5) or on the hind legs (Figure 1-6). The pollen basket is a concave area surrounded by hairs on the hind legs (Figure 1-7). A few bees carry their pollen internally (Figure 1-8). Only female bees have a scopa or corbicula, as they alone collect pollen for nesting.

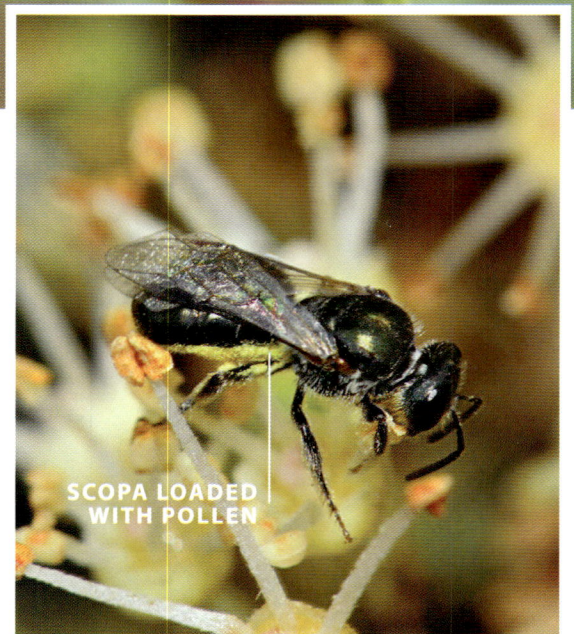

SCOPA LOADED
WITH POLLEN

FIGURE 1-6

In this bee (Halictidae: Halictinae: *Homalictus dampieri*), the scopa is a large area on the underside of the body and the hind legs.
IMAGE **GIORGIO VENTURIERI**

FIGURE 1-7 Workers of highly social stingless bees and honey bees sport a corbicula or pollen basket, a concave area surrounded by hairs on the hind leg.
IMAGES **CAROLE BRYANT** (*LEFT*)
GLENBO CRAIG (*RIGHT*)

FIGURE 1-8 This Hylaeinae bee (*Meroglossa itamuca*) has no scopa or hairs to catch the pollen; instead, she swallows pollen and carries it internally in her crop.
IMAGE **TOBIAS SMITH**

Other changes that help bees to collect food include branched hairs with lots of surface area to collect pollen grains (Figure 1-9), and **brushes** and **combs** on their legs to groom the body and move the scattered pollen to a secure place for transport back to the nest.

FIGURE 1-9 The hairs on a bee's body may be branched like these hairs on a stingless bee, making them well suited for collecting pollen. Wasps and other insects have simple, non-branched hairs, like humans. ELECTRON MICROSCOPE IMAGE **BRONWEN CRIBB**

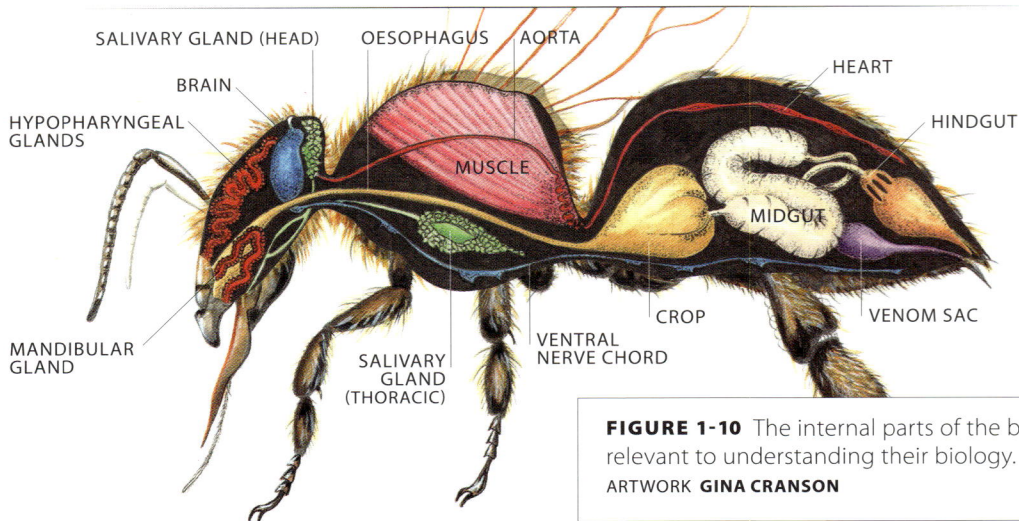

FIGURE 1-10 The internal parts of the bee relevant to understanding their biology.
ARTWORK **GINA CRANSON**

Inside a bee's exoskeleton are its internal body parts (Figure 1-10). A bee's head contains its **brain** and several types of **glands.** These glands have various functions, such as producing digestive enzymes and pheromones. In the social bees, young adult bees called **nurse bees** have active **hypopharyngeal glands** that secrete nutritious brood food, which they deposit into brood cells to nourish immature bees. The thorax is the base for the legs and wings, and is crammed full of muscles required to power these organs of locomotion.

The abdomen contains the important internal organs. For example, the **crop** lies here. Ingested nectar is carried in this sac-like "honey stomach". The crop can expand enormously to allow a foraging bee to carry a full load of nectar back to the nest. The workers in social bees have **wax glands** between the plates that make up the exoskeleton (Figure 1-4). These glands secrete sheets of wax, **beeswax**, which are carried away by nestmates to be used in building nests. The wax glands of honey bees are on the underside of the abdomen, while those of stingless bees are on their top.

In female bees, the tip of the abdomen bears the **sting**, complete with **stinger** and **venom sac**. Like their wasp ancestors, all bees possess a sting. Even in "stingless" bees, the sting is actually present as a tiny, non-functional remnant. The sting is functional in all other bees (97% of species), including the solitary bees. Most solitary bees will sting only if handled. Only honey bees brandish **barbed** stings that remain in the victim and result in the death of the bee.

Bees, like most insects, see a spectrum of colours, but that spectrum is shifted relative to humans (Figure 1-11). Bees see less red but more ultraviolet than we do. Red flowers are usually not strongly attractive to bees (but are to birds). On the other end of the spectrum, bees can see "nectar guide patterns" formed by ultraviolet pigments that direct them to floral resources.

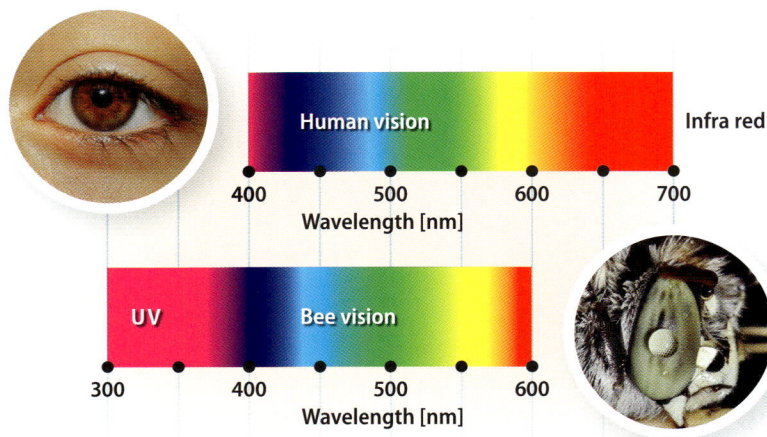

FIGURE 1-11 Bees see less red but more ultraviolet than humans.
BEE EYE IMAGE **LAURENCE SANDERS**

1.4 The bee life cycle

The life of a bee starts when an adult female bee lays an **egg**. The **larva** (Figure 1-12) hatches from the egg, feeds, and grows. The larva does not leave its cell during this time, is legless and is not capable of much movement or any other activity. When the larva reaches its full size, it stops feeding and defecates. (It does not defecate until it has consumed the last of its food or it would foul its food. A constriction at the junction of the **midgut** and **hindgut** prevents defecation until it has finished feeding.)

Many species then spin a **cocoon**. The silk threads that form the cocoon are produced in the **salivary glands** of the larva and secreted through the mouth. Inside the protective cocoon, it changes into a **pupa**. The pupa appears to be inactive but it is undergoing comprehensive changes from a simple larva to a complex **adult**, in a process known as **metamorphosis**.

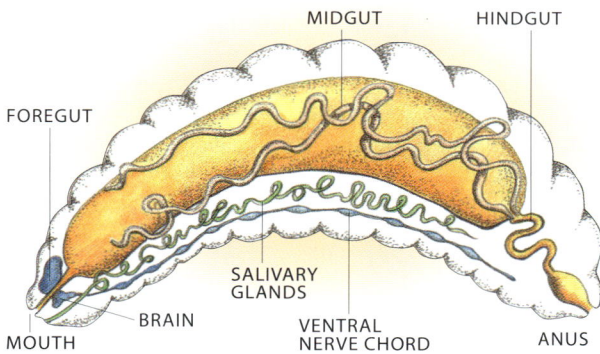

FIGURE 1-12 An internal view of a bee larva.
ARTWORK **GINA CRANSON**

1.5 An introduction to bee social life

More than 90% of bee species are **solitary**. But some species have evolved intricate behavioural interactions with other individuals of their species, which we call **social behaviour**. Definitions vary but, in this book, I use a simplified system of describing bee species as being solitary, semi-social or eusocial.

Female solitary bees rear their young alone. The most basic kind of social behaviour is **nest aggregation**, as exemplified by blue-banded bees. More advanced is **nest sharing,** where nest entrances are shared but each female builds and provisions her own cells within the nest (Figure 1-13). The advantage of a shared nest entrance is improved defence, because one female is usually present to deter intruders. Progressively more complex **semi-social** behaviours have evolved, such as **cooperative brood care,** in which females work together to rear their young. Another step is **reproductive division of labour**, in which some individuals dominate reproduction while others are sterile or sub-fertile, and do more work than the reproductive nestmates. This is evident in a few halictine and allodapine species.

FIGURE 1-13
This bee (*Lasioglossum willsi)* shares a nest entrance with other females of her species.
IMAGE **DIANNE CLARKE**

In the more complex **eusocial behaviour,** we see an overlap of generations: that is, the queen lives long enough to meet her offspring as adults (solitary bees usually die before their offspring emerge as adults). The evolution of **eusociality** is considered to be "one of the major innovations in the history of life" by renowned biologist Edward O. Wilson. Eusociality has evolved also in some wasps and in all the ant and termite species alive on the planet today.

Eusocial behaviour in bees may be primitive or advanced. An example of **primitively eusocial** species are the bumble bees. The most advanced **highly eusocial behaviour** has evolved in the stingless bees (Meliponini) and the honey bees (Apini). In these highly eusocial bees, the queen caste totally dominates reproduction, the queen is morphologically distinct from the worker caste, there is no solitary phase in the life cycle of colonies, and colonies are perennial (they live for many years). They are also the only two groups of bees that make and store harvestable amounts of honey. These concepts are explained in more detail in Chapter 3 "Understanding the highly social bees".

Male and female peacock carpenter bees on leptospermum flowers.
IMAGE **CORINNE JORDAN**

2 Meet the bees

Face of minute *Quasihesma walkeri*, named after Dr Ken Walker, native bee specialist at Museum Victoria.
IMAGE **KEN WALKER**

Although this book focuses on stingless bees, it is edifying to explore bees more broadly. In this chapter, I summarise the evolution of bees and give some examples of the diversity of our native and introduced species. Before turning to the highly social bees in the next chapter, we take a brief tour of some solitary and semi-social species to illustrate their range of forms, life histories and behaviours.

FAST FACTS

Bees appeared 120 million years ago and have diversified into seven families.

The number of bee species in the world, 25,000, is more than the total number of bird and mammal species combined.

The 2,000 species of Australian native bees show a huge diversity of size, shape, nesting biology, behaviour, and social complexity.

Cleptoparasitic bees do not build their own nests, but lay eggs within the nest of their hosts.

Some solitary bees can be fostered by providing them with nesting materials.

Six bee species have been introduced to Australia, including the important honey bee.

The bee tree

WASPS

BEES

MELLITIDAE

ANDRENIDAE

HALICTIDAE

STENOTRITIDAE

COLLETIDAE

MEGACHILIDAE

APIDAE

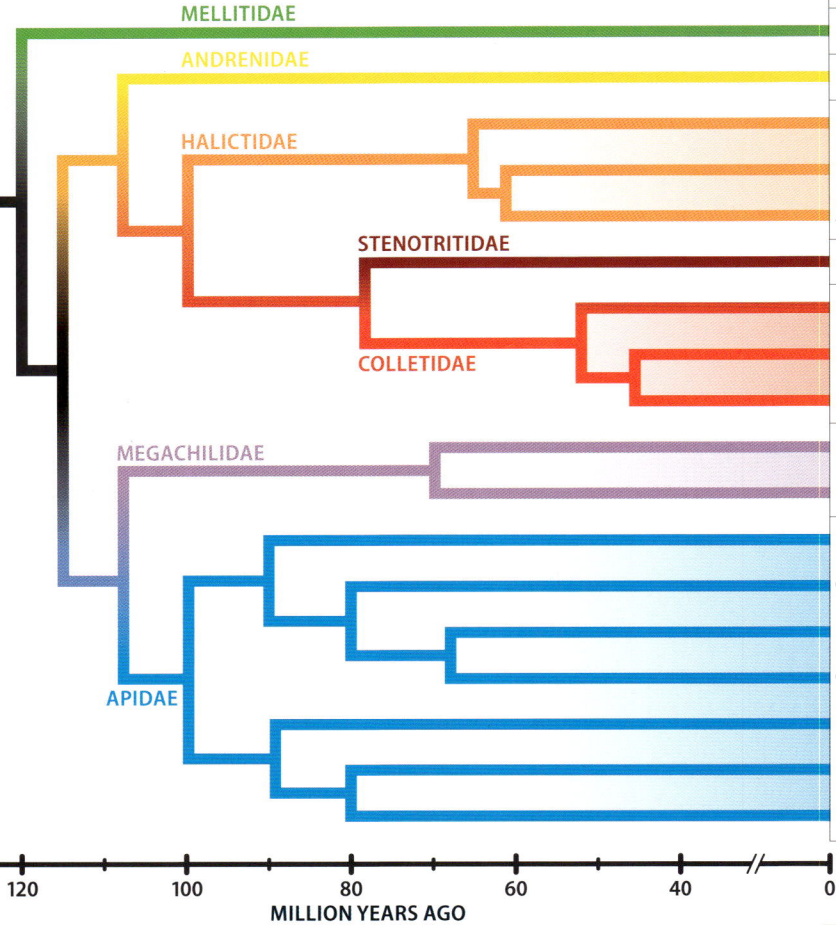

		Family, subfamily or tribe	Examples
			(NOT IN AUSTRALIA)
			(NOT IN AUSTRALIA)
		Nomiinae	*Lipotriches australica*
		Nomioidinae	-
▲	○	Halictinae	*Homalictus dampieri*
		Stenotritidae	*Ctenocolletes nicholsoni*
		Colletinae	*Leioproctus incanescens*
		Euryglossinae	*Pachyprosopis eucyrta*
■	○	Hylaeinae	*Meroglossa itamuca*
		Lithurginae	-
	○	Megachilinae	*Megachile deanei*
▲	○	Xylocopinae	*Xylocopa lieftincki*
✳		Apini	*Apis mellifera*
△		Bombini	*Bombus terrestris*
✳		Meliponini	*Tetragonula carbonaria*
		Anthophorini	*Amegilla murrayensis*
●		Melectini	*Thyreus nitidulus*
●		Nomadinae	-

120   100   80   60   40   0
MILLION YEARS AGO

■ **Semisocial** *(some species)*
▲ **Eusocial, primitive** *(some species)*
△ **Eusocial, primitive** *(all species)*
✳ **Eusocial, advanced** *(all species)*

● **Parasitic** *(all species)*
○ **Parasitic** *(some species)*

FIGURE 2-1
This phylogenetic tree shows the major groups of bees placed on branches that represent natural evolutionary pathways. The estimated time at which branching occurred is on the bottom axis.
ARTWORK GLENBO CRAIG

2.1 The evolution of bees

Ancestral bees appeared about 120 million years ago. Over millions of years, they have evolved into a huge diversity of species. Today, approximately 20,000 species have been named, and this may reach 30,000 once the enormous task of describing and cataloguing them has been completed. This gives bees more species than birds and mammals combined. Bees have spread to most parts of the world. In Australia alone, 1,660 species have been named, with an estimated total of around 2,000 species.

Like all animals, bees are arranged into **families,** and the families are further divided into **sub-families**, **tribes**, **genera** and **species**. So, for example, the species *Tetragonula carbonaria* belongs to the family Apidae, sub-family Apinae, tribe Meliponini, genus *Tetragonula*, with the species name *carbonaria*.

Examples

FIGURE 2-2
An example of a halictid bee in the sub-family Nomiinae, *Lipotriches* sp., a common native bee that nests in tunnels it constructs in soil.
IMAGE **TOBIAS SMITH**

We can represent this diversity on a **phylogenetic tree** (Figure 2-1), which reflects natural evolutionary pathways to the best of our current knowledge. The seven families of bees are divided into the long-tongued bees (two families: Apidae and Megachilidae) and the short-tongued bees (the rest). The make-up of the Australian fauna differs from elsewhere in that it is dominated by the primitive short-tongued families Colletidae and Halictidae. Examples from most of the major groups of bees occurring in Australia are included in this phylogeny and in the text.

In addition to showing the evolutionary pathways and relations between groups, the phylogenetic tree can also show extra information. Various levels of social behaviour are revealed by symbols (Figure 2-1). It is apparent that social behaviour has evolved independently numerous times, as shown by its appearance at several points on the phylogeny. The groups in which parasitic behaviour has evolved are also indicated.

This phylogeny is based on the work of many scientists over nearly three centuries, including the great bee diversity specialist, Charles Michener. The new generation of evolutionary biologists, such as Bryan Danforth, uses molecular techniques to suggest modifications to the structure and to estimate time scale (Figure 2-1).

2.2 Some dazzling examples of native solitary bees

Five of the seven families of bees occur in Australia. One of them occurs only in Australia, nowhere else on the planet. To give a taste of the diversity, I give examples from each of the families and sub-families that we are privileged to share a continent with. They illustrate the diversity of size, appearance, nesting habits and foraging behaviour.

These examples cover the spectrum from solitary species to the highly eusocial stingless bees. There is something for everyone here!

Bees of the large family **Halictidae** (Figure 2-1) nest in soil and forage in natural landscapes, gardens and agricultural areas, where they can sometimes be very abundant.

Bees of this family are often considered to be great pollinators because they carry their pollen loosely scattered on their hairy legs and underside of the body (Figure 1-6, Figure 2-2), increasing the probability of transferring some of this pollen from flower to flower. Many species are also buzz pollinators (read more about this later in this section). Halictids show a wide range of social behaviours. Several species are parasites on other species of halictid bees.

Australia is the only home to a family of bees known as the **Stenotritidae**, a small family of only 21 species. These bees are fast-flying, ground-nesting, solitary, large and hairy (Figure 2-3).

About half of all native Australian bee species belong to the **Colletidae** (Figure 2-1), a huge family of solitary bees. In Chapter 1, we met *Meroglossa itamuca* (**Hylaeinae**), which lacks a scopa and instead carries pollen exclusively in its crop (Figure 1-8).

Another large sub-family, the **Euryglossinae,** also carries pollen internally. Species of Euryglossinae are small to minute, ground-nesting bees that occur only in Australia. They are often found on the flowers of Myrtaceae, that most typical family of Australian plants, which includes eucalypts, paperbarks and tea-trees. Although these bees are very small, close-up pictures reveal their glorious beauty (Figure 2-4).

A striking Australian example of the **Colletinae** are the persoonia bees *Leioproctus (Cladocerapis)*, which specialise in feeding on flowers of *Persoonia* (commonly known as "geebungs"). The face and front legs of these bees are modified so that the bees can gain access to hidden pollen (Figure 2-5). In this exceptional example of coevolution between plant and insect, the parties depend on each other: the bees derive their food only from *Persoonia*, and the plant can only be pollinated by these bees.

FIGURE 2-4 Examples of Euryglossinae bees.
A. *Pachyprosopis eucyrta* from Western Australia, on a eucalypt flower. IMAGE **JEAN AND FRED HORT**
B. *Brachyhesma houstoni.* IMAGE **JEAN AND FRED HORT**
C. *Euryglossa depressa.* IMAGE **REINER RICHTER**

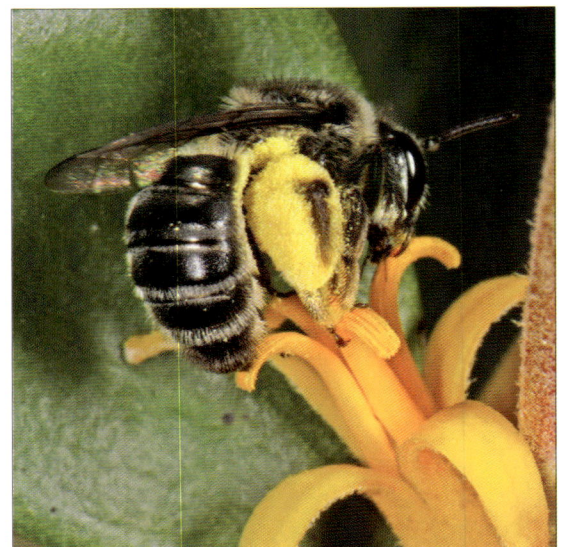

FIGURE 2-3 An example of the small Australian family, the Stenotritidae, *Ctenocolletes nicholsoni* foraging on Scaevola flowers in Western Australia.
IMAGE **LINDA ROGAN**

FIGURE 2-5 An example of a bee in the sub-family Colletinae, *Leioproctus (Cladocerapis) incanescens.* This bee is an ecological specialist that only visits flowers of one type of plant. IMAGE **MARC NEWMAN**

Australia is home to about 150 species of bees in the family **Megachilidae** (Figure 2-1). These bees carry their pollen on the underside of the abdomen (Figure 1-5). They collect materials such as leaf discs (leaf pieces cut using their mandibles) or resin (a sticky substance from plants) to line their nests. **Leaf-cutter bees** are often seen building nests in various holes and spaces around the garden. They are less often seen cutting their characteristic shapes from soft leaves, because they work very quickly. The resulting damage to leaves is often seen in gardens (Figure 2-6, Figure 2-7). Their nests are marvellous structures of leaves where the young develop (Figure 2-8, Figure 2-9).

FIGURE 2-6 Damage to leaves caused by leaf-cutter bees; females cut leaf discs to line their nests.

FIGURE 2-7 A female leaf-cutter using her mandibles to cut a leaf disc, which she folds between her legs for transport back to her nest.
IMAGE **TOBIAS SMITH**

FIGURE 2-9 This larval resin bee has consumed its provisions and is ready to pupate in its nest (which has been opened to photograph); it will later emerge as an adult to complete the life cycle.
IMAGE **GLENBO CRAIG**

FIGURE 2-8 A leaf-cutter bee nest consisting of a series of cells within a hollow stem. Each cell is provisioned with pollen and nectar. The adult female then lays an egg and uses more leaf discs to close the cell and provide a safe haven for her young to develop. IMAGE **GLENBO CRAIG**

Resin bees line their nests not with leaves but with plant resins combined with chewed leaf material (Figure 2-10). You will sometimes see them around remnants of stingless bee nests (Figure 2-11). Their nests are sticky structures but safe refuges for their young (Figure 2-12).

FIGURE 2-11 A resin bee (*Megachile deanei*) scavenging for resin from an abandoned stingless bee nest. IMAGE **CORINNE JORDAN**

FIGURE 2-10 Golden-browed Resin Bee (*Megachile aurifrons*) at its nest entrance.
IMAGE **TOBIAS SMITH**

FIGURE 2-12 Larva of a resin bee feeding on the provisions in its resinous nest (which has been opened to photograph). IMAGE **ERICA SIEGEL**

The **Apidae** contains an impressive diversity of bees (Figure 2-1). In Australia, the **Apini** include two species of introduced *Apis* honey bees (Figure 2-23, Figure 2-24) and the **Bombini** is represented by just one species of introduced bumble bee in Tasmania (Figure 2-25).

The **Meliponini** are the stingless bees that are the central characters of this book. **Xylocopinae** is represented by reed bees (e.g. *Exoneura*) (Figure 2-29) and large carpenter bees of the *Xylocopa* genus (Figure 2-13). Carpenter bees are so named because they drill their nests in wood (Figure 2-14). In some parts of the world they are considered pests because they damage timber buildings but, in Australia, they normally nest in soft woody plants or pithy stems (Figure 2-15).

FIGURE 2-13 Carpenter bees are among our most conspicuous species, with their large size, loud flight and bright colours. Above is a great carpenter bee *Xylocopa (Koptortosoma) lieftincki*. IMAGE **TOBIAS SMITH**

FIGURE 2-14 This peacock carpenter bee *Xylocopa (Lestis) aeratus* is peering out from the entrance hole of a tunnel drilled in timber. IMAGE **REMKO LEIJS**

Insert image shows a *Xylocopa (Lestis) bombylans* at the entrance to a tunnel. IMAGE **ERICA SIEGEL**

FIGURE 2-15 BELOW
Brood cells in the nest of a carpenter bee *(Xylocopa (Lestis) aeratus)*. Nest entrance is on the left. The four cells from the left are of increasing age. The first cell contains a larva eating provisions. In the second cell the egg may have died as the provisions appear untouched. The final two cells contain larvae that have consumed their food. The walls between each cell are constructed from chewed timber, a by-product of the tunnelling. IMAGE **REMKO LEIJS**

Carpenter bees display a behaviour called **buzz pollinating**, or **sonication**. They take their wings "out of gear" and "rev the engine" (contract their flight muscles) to vibrate their entire body while holding tightly to a flower. The vibrations release pollen from the plant's anthers (Figure 2-16), providing the bee with food and the flower with a pollination service.

FIGURE 2-16 This carpenter bee is buzz pollinating a melastoma flower. Note the pollen being sprayed out by the vibrations.
IMAGE **CORINNE JORDAN**

Another buzz pollinator is the **blue-banded bee**, *Amegilla* spp., which belongs to the **Anthophorini** (Figure 2-1). These gorgeous and useful insects are highly adaptable to disturbed human-modified landscapes, and are one of the most common species of native bees seen in Australian gardens. They are solitary, with each female building her own nest. But they nest gregariously, meaning that many females build their nests close together. The natural nest site is soft, dry soil (Figure 2-17). They often enjoy the cool and dry soil conditions under older style houses built on stumps. They can become a pest of mud-brick or rammed-earth houses and can even excavate the mortar between fired bricks. But they can be encouraged to nest locally and harmlessly by providing mud structures of a suitable consistency (Figure 2-21).

FIGURE 2-17 A blue-banded bee, *Amegilla* sp., returning to its nest tunnel with hind legs covered in pollen. IMAGE **TOBIAS SMITH**

FIGURE 2-20 A cuckoo bee *(Thyreus nitidulus)* feeding on nectar from a flower. These bees parasitise the nests of their host bees. IMAGE **DIANNE CLARKE**

FIGURE 2-18 A group of male blue-banded bees clustering at dusk on a stem. The top two males show how these bees hold the stem with their mandibles, not their legs. IMAGE **TOBIAS SMITH**

Male blue-banded bees can be seen roosting near nests in the evening (Figure 2-18). Like all male bees, they do not build nests. Also in common with all male bees, they do not have a sting.

A substantial proportion of the flowers on our planet require buzz pollination. Many members of the plant family Solanaceae require buzzing. Even some of our crops, such as tomatoes, need buzzing when grown in greenhouses (Figure 2-19). (Read more about managing sonicating native bees in Chapter 15 "Bees and pollination".)

The **Nomadinae** and **Melectini** (Figure 2-1) are **cleptoparasitic bees,** also known as "**cuckoo bees**". Females of about 15% of the world's bees do not collect pollen nor build nests but parasitise the nests of other honest and hard-working species. In cleptoparasitism, or parasitism by theft, the adult female infiltrates the nest of its pollen-collecting host, lays an egg in a provisioned cell, and leaves. The cleptoparasite egg develops on the nest provisions of the host. Cleptoparasitic bees are common around bee hotels and are obvious because of their stunning beauty (Figure 2-20).

Cleptoparasitic bees lack a scopa for carrying pollen. They visit flowers only for the energy-boosting nectar. They are usually well armoured for protection against the sting of their host.

2.3 Bee hotels and gardens

Who needs a hive? Solitary bees don't, but you can provide them with artificial nest sites (Figure 2-21). An initiative that is gaining popularity globally is to offer homes for bees in structures called bee hotels, also known as bee walls or, romantically, as bee love motels!

Solitary native bees naturally nest in soil, hollow stems, pithy stems, solid wood, decaying wood, burrows in wood, etc. You can attract a diversity of local solitary bees by furnishing your hotel with various nesting materials, such as mud bricks, bunches of hollow stems, paper tubes, drilled wood blocks, etc. For an excellent guide

FIGURE 2-19 The blue-banded bee *(Amegilla murrayensis)* buzz pollinating a tomato flower. IMAGE **KATJA HOGENDOORN**

to producing bee hotels, see Glenbo Craig's *Planting and Creating Habitat to Attract Bees*, available for free from the website of the Mary River Catchment Coordinating Committee (www.mrccc.org.au).

Bee hotels may boost bee populations and crop pollination on farms. They are also a wonderful addition to a garden, being entertaining and a great educational tool. But take care! First, the hotel will need protection from marauding ants from below and rain from above. Second, be prepared for crowds of wasps to also check into the hotel. (Most of these wasps are beneficial natural enemies of plant pests, so make them welcome too.) Third, bee hotels can be dominated by non-native bees; in Australia, the only invader we have that uses hotels is the African carder bee (Figure 2-27). Fourth, there is little evidence that bee hotels actually boost local bee populations. Fifth, bee ecologist David Roubik warns that bee hotels and the resulting aggregation of bee nests can be magnets for predators and parasites (including cleptoparasitic bees) and help bee diseases to multiply. So, a more successful tactic could be to disperse nesting materials, rather than concentrating them in one position.

FIGURE 2-21 An example of a bee hotel at Gympie Landcare Nursery in Queensland. IMAGE **GLENBO CRAIG**

An easier option than building a hotel is to leave a little wild space in your garden. When you are next thinking about cleaning up that untidy part of your yard, don't do it! Go for a walk instead and leave that refuge for the small creatures. Don't mow that patch of native grass but leave it to shelter the ground-nesters among the tussocks. Leave that dead tree in place to decompose and provide generations of wood-boring bees and

FIGURE 2-22 A bee-friendly garden.
IMAGE **JERRY COLEBY-WILLIAMS**

other insects. Pile up garden debris in a hidden corner and let it form a sanctuary for an insect menagerie.

If you do build a bee hotel, be aware that it works best if married to a bee-friendly garden. Differing motivations and interests generate distinct "bee garden" styles, but all can provide excellent bee forage. Some horticulturists are stimulated by a desire to grow their own food locally, sustainably and organically. These permaculture gardens satisfy a number of fundamental human needs, as well as the needs of bees and, in turn, the bees help with pollination. Other gardeners are driven by a passion to restore natural ecosystems and create native gardens that are havens for wild plants and animals, including bees. A third style of garden is dominated by ornamental plants that enrich our lives with their colour, texture, form and scent. Jerry Coleby-Williams has combined an ornamental and kitchen garden to create a bee haven on his small plot of land in suburban Brisbane, where he has catalogued 26 species of bees (Figure 2-22).

Be aware that, for prolific bee populations, it is not so important what you plant on your plot but what is available in the wider environment. This is because bees will forage over several hundred metres or even some kilometres, so can call on flowers over a large area. However, you can observe and enjoy bees more frequently in your garden if you grow species that are attractive to them.

FIGURE 2-23 A European honey bee foraging on a poinsettia flower. IMAGE **GLENBO CRAIG**

2.4 Bees introduced to Australia

Australia is relatively free of introduced bee species. Of the six species that have naturalised since European settlement, two were deliberately and legally released, while the remainder have invaded by their own means.

2.4.1 EUROPEAN HONEY BEE (Apis mellifera)

The **European honey bee** was introduced into Australia in 1822 by English settlers. *Apis mellifera* is also called the **Western honey bee** or **common honey bee.** I call it the European honey bee in this book, not because the species is restricted to Europe (it also occurs in Africa), but because Europe is where the races of this species used for beekeeping come from. Nearly half a million colonies are now managed in hives in Australia. These hives form the basis of an important industry that produces honey (worth about $100 million per year) and provides crop pollination services (valued at between $4 billion and $6 billion annually). European honey bees visit a huge number of native and exotic flowers (Figure 2-23).

This species is an extremely successful invader, with colonies escaping into the wild by reproductive swarming or absconding. These feral colonies are abundant over most of the Australian continent, except for inland areas away from water. The environmental impacts of honey bees are not well studied, but there is evidence that, in some situations, they out-compete native insects and birds for floral resources and nest sites, reduce seed set of native plants, and increase the seed set of weeds. In New South Wales, they are listed as

a Key Threatening Process. In South Australia and Western Australia, feral hives are destroyed to protect rare cockatoos. You will hear a lot more about the European honey bee as you make your way through this book.

2.4.2 ASIAN HONEY BEE (Apis cerana)

The Asian honey bee entered northern Australia through the port of Cairns in 2007. An eradication campaign failed and efforts now focus on containing its spread and impact. It has a native range throughout Asia and, like the European honey bee, includes various races of bees.

Like the other 10 honey bee species, the Asian honey bee is highly social. Individual bees (Figure 2-24) are smaller than the European honey bee, as is the number of bees in the colony. Colonies can use smaller cavities for nesting than European honey bees. They are hard to manage in hives as they abscond (abandon their nests) readily in response to a food shortage, attack or disease outbreak. They are timid in defence but can raid the nests of European honey bees to steal stored food (the opposite can occur too).

The value of Asian honey bees as managed pollinators is lessened by the fact that they tend to forage from numerous minor sources of nectar and pollen rather than from a major source such as farm crops. On the other hand, they do have the advantage that they are resistant to the varroa mite (*Varroa destructor*), a pest that has decimated European honey bee populations in some countries. Asian honey bees are the original host for varroa mites, and concern has been expressed that they may carry the mites into Australia,

FIGURE 2-24 A foraging Asian honey bee in far north Queensland. IMAGE **TOBIAS SMITH**

FIGURE 2-25 A foraging bumble bee in Tasmania. IMAGE **TOBIAS SMITH**

where they could attack European honey bees. This is unlikely to be an immediate problem, as the races of varroa hosted by Asian honey bees would not be adapted to attack European honey bees. However, the mites could adapt over time to attack European honey bees, as occurred in Asia. If this did happen, Asian honey bees, with their resistance to varroa mites, could potentially fill the gap of free, incidental pollination currently provided by feral European honey bees.

That said, these bees do have a downside. If they spread across Australia, they could be expected to reduce the amounts of nectar and pollen available to birds and native insects. They are also likely to forage further inside rainforest than European honeybees. Because they will nest in small cavities, they have stung people in Cairns after emerging from letter boxes and small holes in external walls.

2.4.3 BUMBLE BEE *(Bombus terrestris)*

The world has about 250 species of bumble bee but none occur naturally in Australia. All bumble bees are primitively eusocial species: they show overlap of generations but the female castes are flexible (the nest is founded by an individual that starts as a worker and then becomes a queen), colonies are founded by a single female, and the colonies are annual not perennial (they die out at the end of each season and are founded afresh at the beginning of the following season).

Bombus terrestris is the largest and one of the most abundant bumble bee species in its native Europe. It has been developed as a managed pollinator on a very large scale, with global trade of around one million colonies per year. It has been introduced to Japan, the USA, Chile, New Zealand and elsewhere for pollination, particularly in greenhouses.

Bumble bees cannot be legally imported to Australia but were introduced (illegally or accidentally) to Tasmania in 1992, where they quickly established over a wide area. *Bombus terrestris* has a record of successful invasions in other parts of the world, with consequences for native bees, native plants species and invasive weeds. This species has a broad diet and forages on both native and introduced plants (Figure 2-25). It is an effective pollinator of many weeds, some of which have shown increases in Tasmania attributed to bumble bee pollination. Hence it is likely that it will have a negative impact on the Australian environment. Bumble bees are prohibited from being moved from Tasmania to other states or territories, because of the impact they are likely to have by competing with birds and insects and by pollinating weeds. Even in Tasmania, their use for glasshouse pollination is prohibited.

FIGURE 2-26 Alfalfa leaf-cutter bee on lucerne flowers. IMAGE **DENIS ANDERSON**

FIGURE 2-27 An introduced adult African carder bee collecting nest material. IMAGE **DIANNE CLARKE**

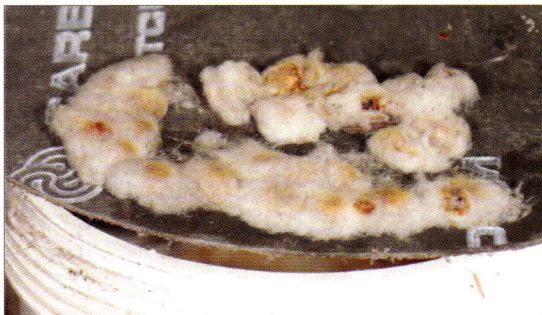

FIGURE 2-28 A nest of introduced African carder bees in an electricity meter box. IMAGE **JAN ANDERSON**

2.4.4 ALFALFA LEAF-CUTTER BEE
(*Megachile rotundata*)

The alfalfa leaf-cutter bee is a solitary (but gregarious) bee of European origin. It was deliberately imported into North America for pollination of alfalfa (lucerne) and other crops. It is a very efficient pollinator of lucerne (Figure 2-26) and is managed on a large scale and at a high level of sophistication for crop pollination. It was legally introduced into Australia from 1987. It has not prospered in the Australian environment and certainly has not been a commercial success. Agronomist James De Barro states that a small unmanaged population exists in the Keith area of South Australia.

2.4.5 THE EMERALD FURROW BEE
(*Seladonia hotoni*)

This small metallic green bee was first discovered in Australia in 2004 and is now known from the Sydney Basin and Hunter Valley. It was first identified as *Halictus smaragdulus* but investigations by Michael Batley and colleagues revealed its true identity. It is native to the southern Africa, where it nests in soil and so might have been introduced into Australia in nesting tunnels in soil. It is commonly seen on flowers of the introduced species *Galenia pubescens* (Aizoaceae). Eradication of this species was not attempted in Australia as it has no obvious detrimental impacts on the environment.

2.4.6 AFRICAN CARDER BEE OR METER-BOX
CARDER BEE (*Afranthidium repetitums*)

This bee was first detected in Australia in 2000 and, by 2015, was known from Rockhampton to Melbourne. It seems to prefer suburban gardens. It is a solitary bee in the sense that each female makes its own nest, but it does nest gregariously with other solitary females in the same locale. Nest sites include artificial recesses such as electricity meter boxes and sliding window tracks. Females line their nests with plant hairs, which they scrape from leaves with their mandibles (Figure 2-27). This nesting material resembles yellowish cotton wool (Figure 2-28) and cannot be confused with the nesting material of any native species.

In this chapter we have sampled the diversity of Australia's native bees, most of which are solitary. In the next chapter we cover the highly social bees. The evolutionary steps that lead to social behaviour are apparent in the semi-social species, such as this species, *Exoneura bicolor*, an allodapine bee from the subfamily Xylocopinae. Females of these bees share a nest, cooperate in maintaining the nest and foraging, and share egg laying more or less equally.

FIGURE 2-29 (top) *Exoneura* adult and larvae in hollow stem nest. IMAGE **TOBIAS SMITH**

FIGURE 2-30 *Exoneura bicolor* female foraging. IMAGE **LINDA ROGAN**

3 Understanding the highly social bees

Honey bees and stingless bees display the most **advanced social behaviour** of all bees.

The three castes in each colony have defined roles: the queen lays all the eggs, the male drones fertilise the queens, and the sterile female workers do all the work.

Just like a worker bee, **the queen** comes from a fertilised egg, but she is reared in a larger cell and gets more food (and better food, in the case of honey bees).

Drones (males) arise from unfertilised eggs, so they inherit all their genes from the queen.

As workers age, they pass through stages in which they do different jobs, first as house bees in the nest and then as foragers.

Although honey bees and stingless bees **share advanced social behaviour**, it differs in fundamental ways.

Brood comb of the stingless bee *Tetragonula carbonaria*.
IMAGE **GLENBO CRAIG**

3.1 The social life of honey bees and stingless bees

Most species of bees are solitary or semi-social, but two tribes, the honey bees and the stingless bees, have reached a pinnacle of social complexity: they are **highly eusocial** (truly social).

The stingless bees (tribe Meliponini) and the honey bees (tribe Apini) are related but they separated from each other around 80 million years ago (Figure 2-1). Since then, stingless bees have journeyed to their current distribution of all tropical parts of the world. The honey bees now occupy the Old World and never naturally reached Australia or the Americas.

The common ancestor of these two tribes was probably a **primitively eusocial** bee, like a modern bumble bee (Bombini, Figure 2-1). Primitively eusocial bees display overlap of generations and the females occur as castes, but the castes are flexible, colonies can be founded by a single female, and the colonies are annual, not perennial.

Honey bees[1] and stingless bees evolved their highly social behaviour independently on different sides of the globe, so they differ in many respects. But they do have some significant aspects in common. These highly social bees 1) have **female castes**, 2) **display no solitary phase in the life cycle of their colonies** and 3) **create perennial colonies**.

In stingless bees and honey bees, the two female **castes** are the reproductive **queen** and the non-reproductive **worker** (Figure 3-1). The queen caste totally dominates reproduction and is morphologically distinct from the worker caste. The castes are inflexible, that is, a female cannot change her caste as an adult. If a female egg is laid into a normal-sized worker cell, then the resulting larva will develop into a worker and will remain a worker. If a female egg is laid

into a larger queen cell, it will develop into a queen, which also cannot change (Figure 3-2). Males constitute a third caste: the **drones**.

No solitary phase in the life cycle of colonies means that colonies cannot be initiated by a single individual, but require a group of bees working cooperatively. Colonies have their own life cycle, and are born (founded), live, and die. (Read more about this in "Founding a new colony" in Chapter 4.) Colonies of the highly social bee species can live many years and so are **perennial**. Perennial life demands that colonies store food for the lean times. Indeed, both stingless bees and honey bees accumulate honey in volumes that are harvestable by humans.

FIGURE 3-1 Queen and worker castes cooperating to rear brood in stingless bee nests; *Austroplebeia cincta* (ABOVE), *Tetragonula carbonaria* (RIGHT).
IMAGE ABOVE **ANNE DOLLIN**
IMAGE RIGHT **JAMES DOREY**

FIGURE 3-2 Brood cells, each containing an immature bee, in a nest of the stingless bee *T. carbonaria* (LEFT) and *T. hockingsi* (RIGHT). The larger open cells on the left will each produce a virgin queen, while other cells, including the smaller open cells on the right, will produce workers.

---

1 Please note, when comparing honey bees and stingless bees in this chapter and the next, I refer mainly to the European honey bee *Apis mellifera*; some of the details may be different for the Asian species of *Apis*.

3.2 Castes

Among both honey bees and stingless bees, members of each of the three castes differ in many aspects, including their abundance in the colony, the roles they perform, and their life span (Table 3-1). Only one **mated queen** is tolerated in a colony but, in colonies of stingless bees, new **virgin queen**s are produced constantly, or nearly so, and virgins routinely coexist with the mated queen. The mated queen is also known as the fertilised queen or reigning queen and virgin queens are also called gynes or even "princesses".

Workers are abundant; I estimate the adult worker population of a colony of *T. carbonaria* to be around 10,000. Megan Halcroft estimates an average of 5,000 for *Austroplebeia australis*. Honey bee colonies are more populous, with up to 50,000 workers. These figures are, of course, highly variable. **Drone** abundance is especially variable for two reasons: first, because they are produced seasonally and, second, because they may leave the colony at various times. In general, drones are present in the hundreds rather than the thousands. Among stingless bees, the brood cells (cells for eggs, larvae and pupae) of males and workers are identical; among honey bees, male cells are wider and stand taller.

Among both honey bees and stingless bees, the queen is responsible for laying all the eggs required by the colony, does no other work, and even lacks the baskets on her legs needed to collect pollen. All workers and drones in the colony are the offspring of the queen: they are all brothers and sisters. The workers toil day and night at the tasks required by the colony. The males' only job is to mate with the virgin queen of another colony (Figure 3-3). There is intense competition for this job, as there is only one queen per colony, which lives for a year or more and mates only once. An average **life span** for workers of approximately 100 days is typical for many species of stingless bees (although *Austroplebeia australis* lives much longer). The life span of males is harder to determine as they usually leave, or are evicted from, the nest before dying.

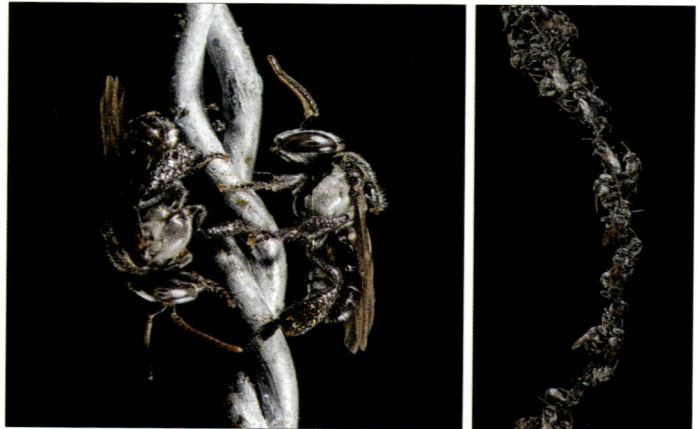

FIGURE 3-3 A congregation of stingless bee males (*T. hockingsi*) roosting at night; these males may be preparing to compete to mate with a queen.
IMAGES **TOBIAS SMITH**

TABLE 3-1 The arrangement of gender and caste with their functions in the social honey bees and stingless bees, with approximate longevities for the Australian native stingless bee *Tetragonula carbonaria*.

Gender	Female	Female	Male
Caste	Queen	Worker	Drone
Abundance	1 mated queen	1000s	100s
Larval cells	Large	Normal	Normal
Tasks	Egg laying	Most jobs in the nest	Fertilise a queen
Life span	Months to years	Approx. 100 days	Short (unknown)

WORKER CELLS

QUEEN CELL

Immature bees (known as "brood") develop in brood cells, progressing from egg to larvae to pupae, then hatching as an adult. In honey bees, the developmental period (the time from egg to adult) differs between castes, with queens, workers and drones developing in approximately 16, 21 and 24 days respectively. In stingless bees, the **developmental period** does not differ between workers and males, and is 50 days for *T. carbonaria*, and 55 days for *Austroplebeia australis*. Stingless bee queens take a few more days to develop, with larger queen cells remaining after drones and workers in adjacent cells have emerged (Figure 3-4).

The longer developmental time for stingless bee queens surprises keepers of honey bees. Yet, in fact, this should be expected, as larger insects normally take longer to develop than smaller ones. So, if this is the case, why do honey bee queens develop quicker than other castes? This has to do with the different way that these two groups of insects naturally re-queen their colonies. In a honey bee colony, the queen that emerges first is the winner of the race to be the next queen, so evolution has favoured queens that

FIGURE 3-4 Queen cells in a nest of *T. carbonaria* remain a few days after adjacent workers and males have emerged, reflecting the extra time taken for queens to develop. IMAGE **GLENBO CRAIG**

develop quickly. But, in stingless bees, it is not a race. Rather, queens are selected by workers from a pool of available "princesses". (Read more about this in "Queen replacement" below.)

The adult honey bee queen is nourished by direct food transfer from attendant workers, but queens of stingless bees tend to feed more from larval food provisions deposited in cells by workers. The queen eats these provisions just before laying her egg in the cell. In some stingless bee species in other countries, workers lay a special egg, called a trophic egg, into brood cells, and this egg is eaten by the queen as a food source. Trophic eggs are infertile and larger than queen-laid eggs. The workers of *Austroplebeia australis* occasionally lay trophic eggs that are eaten by the queen, but her main source of nourishment is still the cell provisions. (Workers of some honey bee and stingless bee species can also lay reproductive eggs; read more about this in "Worker reproduction" below.)

Tetragonula mellipes

Austroplebeia magna

FIGURE 3-5 The hind leg of a stingless worker bee, showing the wide tibia in the form of a basket (corbicula) for carrying pollen; *Tetragonula mellipes* (TOP), *Austroplebeia magna* (BOTTOM).
IMAGES **KEN WALKER**

FIGURE 3-6 The hind leg of a stingless male bee, showing that it is not developed as a pollen basket; *Tetragonula mellipes* (TOP), *Austroplebeia magna* (BOTTOM).
IMAGES **KEN WALKER**

3.2.1 HOW TO DISTINGUISH CASTES

Unlike male honey bees, male stingless bees are similar to female worker bees in size and appearance. But, in both honey bees and stingless bees, the queen is very different from both other castes.

The three castes of adult stingless bees can be distinguished by a number of differences. The workers have a functional corbicula (pollen basket) on the outside of a section of the hind leg called the tibia. This leg section is adapted for carrying pollen by being wide, concave, smooth, and surrounded by a flange of long hairs (Figure 3-5).

Males are free of the task of collecting pollen and, although the tibia remains widened, it is convex rather than concave. Unlike the leg of a worker, the tibia is more or less evenly hairy all over. In the case of the *Austroplebeia* bees, the tibia of males also reveals pale markings (Figure 3-6). In the case of the *Tetragonula* bees, the tibia of males collected outside the nest may be smeared with resin (Figure 3-7). This resin may serve some function during courtship behaviour.

Stingless bee males possess hard external **genitals** to insert into a queen, which allow them to remain attached to her while in flight. These genitals can be seen partly or fully protruding (Figure 3-7). Males are also distinguished by their larger eyes, weaker mandibles and longer antennae with an additional segment (Figure 3-7, Figure 3-16).

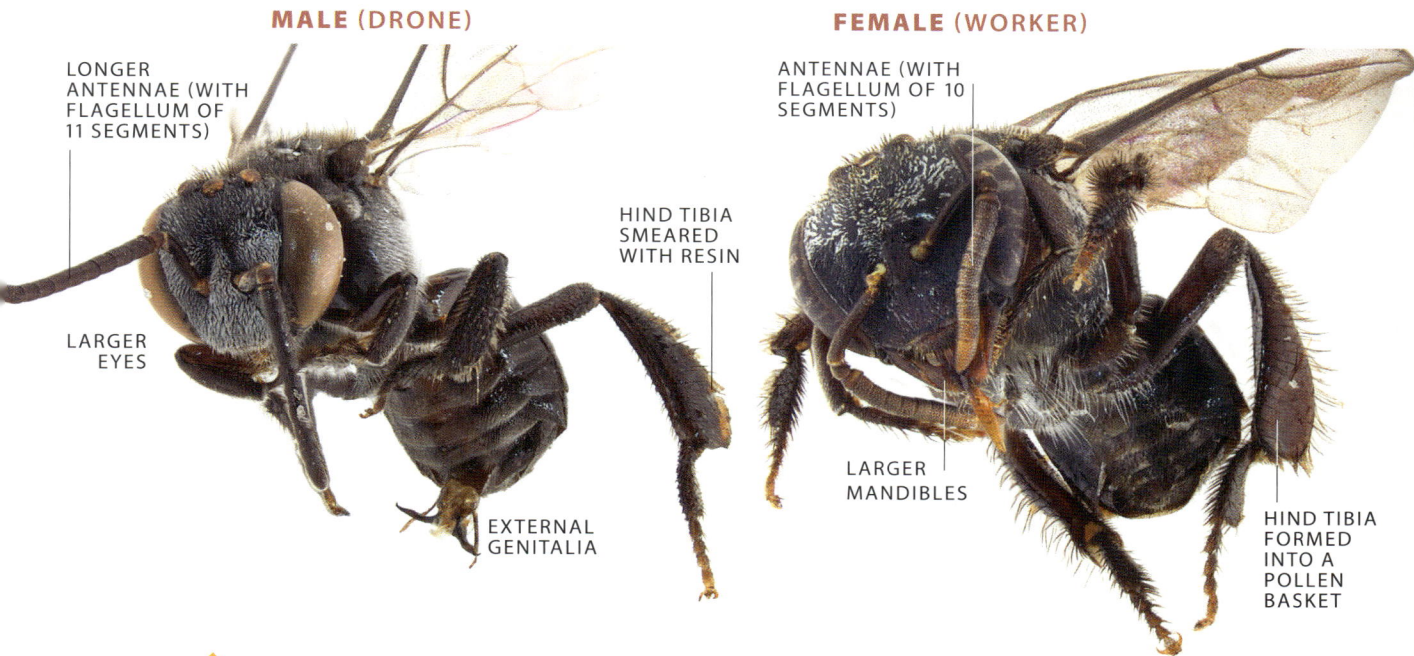

MALE (DRONE)

LONGER ANTENNAE (WITH FLAGELLUM OF 11 SEGMENTS)

LARGER EYES

EXTERNAL GENITALIA

HIND TIBIA SMEARED WITH RESIN

FEMALE (WORKER)

ANTENNAE (WITH FLAGELLUM OF 10 SEGMENTS)

LARGER MANDIBLES

HIND TIBIA FORMED INTO A POLLEN BASKET

FIGURE 3-7 Male *(LEFT)* and worker *(RIGHT)* of the stingless bee *Tetragonula carbonaria*. See the external genitalia, larger eyes, weaker mandibles and longer antennae of the male. IMAGES **JUSTIN BARTLETT**

FEMALE (QUEEN)

FIGURE 3-8 A mated queen of *Tetragonula carbonaria* showing enlarged and striped abdomen. IMAGE **JAMES DOREY**

FIGURE 3-9 A young virgin queen (with three workers) of *Tetragonula carbonaria*, showing the larger abdomen and hairy hind tibia. IMAGE **GIORGIO VENTURIERI**

Both mated and virgin stingless bee queens lack the pollen basket possessed by worker bees (Figure 3-19). The mated queen is larger, with a greatly distended abdomen (Figure 3-20). When the abdomen swells to contain the active ovaries, the segments of the exoskeleton are pushed apart and the pale membranes are exposed, giving the abdomen a striped appearance (Figure 3-8). In this state, the queen is described as **physogastric**. Virgin queens also have a large abdomen, but it is not as distended as that of an egg-laying queen (Figure 3-9).

3.2.2 HOW IS CASTE DETERMINED?

Both honey bees and stingless bees produce new **virgin queens** in special large queen cells. In honey bees, queen production normally only happens in anticipation of swarming, or when a colony is **queenless**. But stingless bees produce queens most of the time, albeit in small numbers (Figure 3-10, Figure 3-11, Figure 3-12). I estimate that only 1 in 1,000 brood cells are queen cells in *T. carbonaria*, which continue to produce queens through all seasons of the year. This is one of the ways that a colony insures itself against loss of the mated queen.

The egg laid in a queen cell is a normal fertilised egg, no different from the eggs laid in worker cells. It is the food that the larva consumes that determines which caste it will develop into. In honey bees, the larvae destined to be queens are fed a larger amount of food and food of a different nutritional make-up (**royal jelly**). (Read more about this in "Brood rearing in social bees" in Chapter 4.) In stingless bees, the only difference is that the larvae destined to be queens are fed larger amounts of food, with the type of food being identical. In both cases, the access to a superior food changes gene expression and stimulates developmental pathways that lead to a queen that is larger, lives longer and behaves in a manner befitting her social role.

FIGURE 3-10 Queen cell of stingless bees (in this case *Tetragonula carbonaria*) is obvious because it is larger and situated on the edge of the comb. IMAGE **JAMES DOREY**

FIGURE 3-11 *(ABOVE LEFT)* A queen cell and three worker cells of *Tetragonula carbonaria* opened to show the mature larvae inside. Note that the larval queen is two to three times larger than other larvae.

FIGURE 3-12 *(ABOVE RIGHT)* The queen cells of *Austroplebeia australis* are more difficult to see in the cluster-like brood of this species, but are still obvious because of their size. IMAGE **JAMES DOREY**

3.2.3 QUEEN REPLACEMENT

Queens of highly social bees do not live forever and need to be replaced, a process called queen replacement or supersedure.

Honey bee workers rear new queens for several reasons: in anticipation of swarming, if the colony becomes queenless, or, in some cases, preemptively, if the mated queen is failing. Honey bee workers construct a number of large queen cells and provision them with royal jelly so the resulting larva develops into a queen. The stinger of the queen is not barbed, so she can sting multiple times without dying, unlike worker bees. She uses the stinger to kill rival queens, either un-emerged queens in cells or other emerged virgin queens.

In contrast, colonies of stingless bees continuously rear a reserve of queens, some of which survive in the colony for extended periods. Stingless bees cannot sting, so queens cannot kill other virgin queen competitors. Instead, when the mated queen fails, the workers select one of the emerging virgins as their new queen, and allow only her to mate. In this way, stingless bee societies resemble a democracy, whereas honey bees behave like a monarchy!

Stingless bee queens can be replaced in a number of ways. Four of these ways are explored on the page opposite.

A · A virgin queen in the colony is recruited to the top job

Virgin queens are usually present in stingless bee nests. They are often secretive, as they may be killed or evicted by workers. But, if the workers perceive that a queen is failing, they will choose a virgin queen and allow her to leave the nest to go on her mating flight. Then they eagerly await her return and accept her back as their new queen.

B · A queen emerges from her cell and is selected to take over

This is very similar to the first mechanism, except that the virgin queen has to hatch from a cell. This may take many days. The queen cells from which these queens emerge are easily seen in the comb-building *Tetragonula* species, but are not so obvious in the species that build cluster broods (compare Figure 3-10 and Figure 3-12).

C · An emergency queen cell is made

Queenless colonies of stingless bees can construct emergency queen cells.

Tulio Nunes and colleagues have recently discovered and documented this for *T. carbonaria*. When a colony is made queenless, it continues to produce brood. These cells contain larval food but not an egg, and many cells are large, similar in size to normal queen cells (Figure 3-13). The position of these large cells is unusual as they are scattered through the comb, unlike normal queen cells, which are always on the periphery of the comb. These are emergency queen cells, and are always built next to a worker cell containing a larva, produced earlier when the colony was **queenright** (i.e. with a laying queen) (Figure 3-14). The worker larva moves into the larger cell, consumes its contents and eventually develops into a queen (Figure 3-15). The process is similar to that of honey bees. Megan Halcroft has observed behaviour in *Austroplebeia australis* colonies that may also be part of an emergency queen-rearing process.

FIGURE 3-13 Brood comb of *Tetragonula carbonaria* one week after queen removal, arrows show larger cells. IMAGE **GIORGIO VENTURIERI**

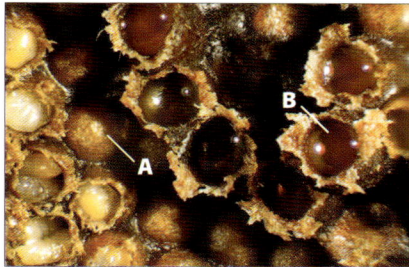

FIGURE 3-14 Brood comb of *Tetragonula carbonaria* two weeks after queen removal, showing (A) the brood produced in queen's presence and (B) brood produced in a queenless state, provisioned with food and not containing eggs. IMAGE **GIORGIO VENTURIERI**

EMERGENCY QUEEN CELL

QUEEN LARVA

EMPTY BROOD CELL

FIGURE 3-15 Queen larva of *Tetragonula carbonaria* adjacent to an empty brood cell. Both cells have been partially opened to show the larva has moved from one cell to the larger adjacent one.

D · A nearby colony contributes a queen

This mechanism is not known for Australian stingless bee species, but a South American species has been shown to do it regularly. A queen from a nearby unrelated nest enters the queenless nest and takes over the role as queen. This is called queen parasitism, as the colony is tricked into accepting an unrelated queen rather than rearing its own queen.

3.3 Gender

As discussed above, stingless bee males have different hind legs lacking a pollen basket (Figure 3-5, Figure 3-6), hardened external genitalia, weaker mandibles, larger eyes and longer antennae than workers (Figure 3-7, Figure 3-16).

In addition to being longer, the male antenna is made up of thirteen segments, one segment more than the twelve segments of females. The antennae are presumably longer to help males locate a virgin queen to mate with. The detection of pheromones by the organs on the antenna will assist this. On the other hand, worker females need more well-developed mandibles to build and defend nests and to forage.

FLAGELLUM OF 10 SEGMENTS IN FEMALE

SCAPE
PEDICEL
FLAGELLUM OF 11 SEGMENTS IN MALE

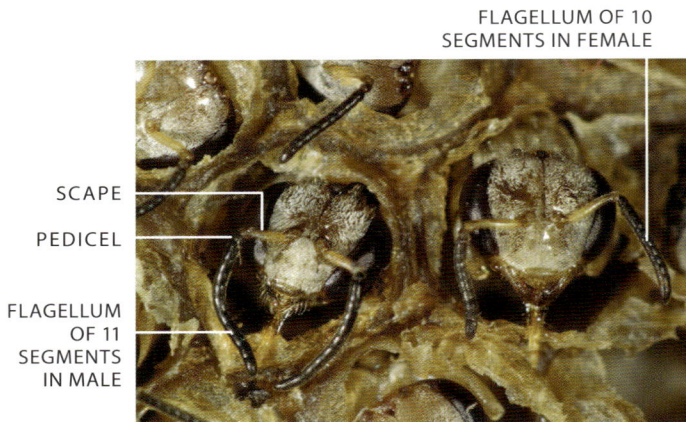

FIGURE 3-16 A male *(LEFT)* and female worker *(RIGHT)* of *Tetragonula carbonaria* stingless bees about to emerge from their natal cells. IMAGE **MALCOLM RICKETTS**

We learnt in the section on caste that colonies constantly produce hundreds of males. Why does a colony produce so many males if they do not do any work and are not even needed to mate with the queen (who chooses a male from another colony)? The answer is that it is a way for a colony to spread its genes. If a male from a colony mates with a queen from another colony, then the genes of the first colony are spread. Colonies with a genetic disposition to produce many males have a greater chance of spreading their genes so this trait will spread.

On the other hand, a colony cannot produce too many males or it will not have enough workers for self-maintenance. The queen determines the sex ratio by controlling the release of sperm and the subsequent fertilisation of the egg.

3.3.1 HOW IS GENDER DETERMINED IN BEES?

The determination of sex in bees (and in their relatives, wasps and ants) is remarkable. The queen lays eggs of both genders. If she fertilises the egg as she lays it, it will become a female. If not, it will become a male. By this means, she can control the sex ratio of her colony (Figure 3-17).

This sex determination system is called **haplodiploidy**, as the females are diploid (with two sets of each chromosome) and the males are haploid (only one copy of each chromosome). Although the queen determines the primary sex ratio, she does not always have the final say, at least in honey bees, because workers determine what is reared by killing males.

Haplodiploidy has some interesting consequences – for example, males have a grandfather but no father!

FIGURE 3-17 *(BELOW)* In all bees and their relatives, a female develops from a fertilised egg, but a male develops from an unfertilised egg.
ARTWORK **GINA CRANSON** AND **GLENBO CRAIG**

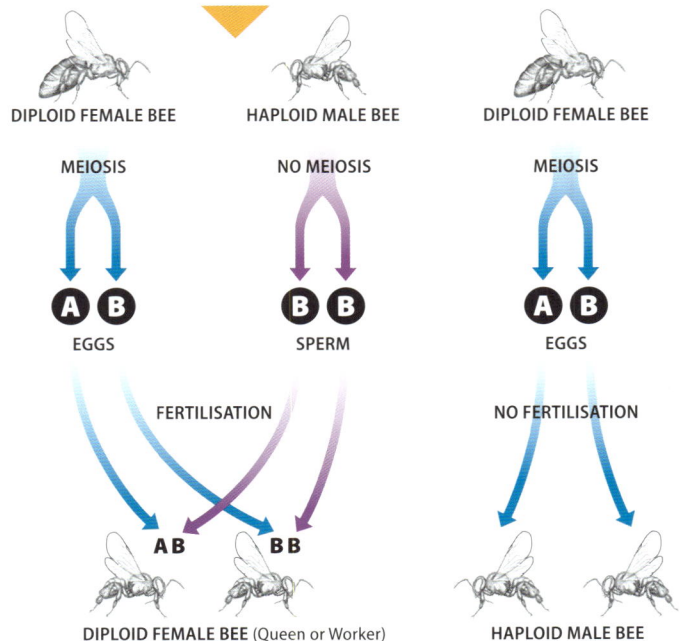

DIPLOID FEMALE BEE   HAPLOID MALE BEE   DIPLOID FEMALE BEE

MEIOSIS   NO MEIOSIS   MEIOSIS

Ⓐ Ⓑ   Ⓑ Ⓑ   Ⓐ Ⓑ
EGGS   SPERM   EGGS

FERTILISATION   NO FERTILISATION

AB   BB

DIPLOID FEMALE BEE (Queen or Worker)   HAPLOID MALE BEE

3.3.2 DIPLOID MALES

If this system of **haplodiploidy** seems curious, then brace yourself, it gets even stranger. Colonies can produce **diploid males** as a result of **inbreeding**. How does this happen?

In the previous section we learnt that sex determination is due to fertilisation of the egg or not. But, when we look at the system in more detail, we learn that it is actually due to a single **sex determiner gene**. If an individual has two different versions of the sex gene (**heterozygous**), the egg will develop into a female. If the egg is unfertilised, then it will be haploid and so have one copy of the gene (**hemizygous**), and will develop as a normal male. However, if an individual's two copies of the gene are the same (**homozygous**), the egg will develop into a male – a diploid male. This happens when a population lacks genetic variation and successive queens mate with their brothers over several generations.

Diploid males are infertile, do not work and impose a drain on the colony. Honey bee workers detect and eat diploid males as larvae, but stingless bees do not appear to do this.

To learn how to manage the risk of diploid males in your bee hives, see "Maintaining genetically viable populations" in Chapter 13.

3.3.3 WORKER REPRODUCTION

In a strange twist, it turns out that the worker caste in some highly eusocial bees – a caste generally regarded as sterile – can also reproduce in a limited way.

Haplodiploidy requires that male bees be produced by parthenogenesis, the development of an unfertilised egg. Workers in the social bees may lay eggs. But the egg is not fertilised because the worker does not have the reproductive system to mate and so does not have sperm available to fertilise the egg. Hence these unfertilised eggs result in males.

This is well known in honey bees, especially when the colony has been queenless for an extended period. Stingless bee workers in some species can also contribute to the production of males in a nest. For example in *Austroplebeia cassiae* and *Austroplebeia australis*, the males are sometimes laid by workers, at least in queenless colonies.

This is not the case in *Tetragonula carbonaria*, in which Ros Gloag and colleagues found that up to 20% of individuals reared in a queenright colony are males but all were laid by the queen. Furthermore, Tulio Nunes and colleagues showed that, even if stingless colonies were kept queenless for weeks, the faithful workers did not start laying eggs.

3.4 Mating

Mating in the highly social honey bees and stingless bees takes place outside the nest, in flight.

A European honey bee queen may take several **mating flights**, close to the beginning of her life as the new queen. A stingless bee queen only ever makes one mating flight. Mating flights occur either when the old queen needs to be replaced, or when a new colony is being founded.

Males from many unrelated colonies form **male congregations** (Figure 3-3). (It is not known from how far away these males come, but their unrelatedness is certainly relevant to maintaining genetic diversity in populations.) The queen flies into the male congregation and mates. Mating is very fast; the male inserts his phallus, ejaculates explosively, drops off leaving part of his phallus in the queen, and immediately dies!

In the stingless bees, the queen will only mate with one male, but a honey bee queen mates with several males in succession during her flight or flights. The queen then returns to the nest and after a few days begins laying eggs. She will not mate again but uses the stored sperm to fertilise all the female eggs she will lay for the rest of her life. The sperm is stored in a special organ called a **spermatheca**, where it is nourished, protected and kept alive. To create a female, the queen releases a few sperm to fertilise the egg as it passes down her reproductive tract. To create a male, she will not fertilise the egg.

Because honey bee queens mate multiple times, their colonies are polyandrous, or many-fathered. Because stingless bee queens mate only once, their colonies are monandrous, or one-fathered. This means that all workers in a stingless bee colony are full siblings, while those in a honey bee colony are mostly half-siblings. These groups of half-siblings are known as "sub-families".

3.5 The super-organism

Insect societies present an extra level of biological organisation: the colony (Table 3-2).

Insect societies can be compared to a higher organism such as a mammal. The colony is like the organism and the insects are like the cells. This is the "super-organism" concept. If a colony has only one queen at any one time and she only mates with one or a few males, only a few individual bees get to reproduce. This is not so different from other animals, in which only a few cells in the body contribute to the next generation. The other cells in our body, such as brain and muscle cells, are like the worker bees in a colony; they help other cells in the organism to reproduce but do not reproduce themselves.

The colony functions like a single organism in many ways. For example, highly social bees are able to maintain stable nest temperatures in the same way that warm-blooded animals do. Decisions can also be made at the colony level (e.g. the decision to found a new colony is made by the whole colony).

The mouth-to-mouth transfer of food from one bee to another is called **trophallaxis,** and is a common and important behaviour in all social insects. Nectar is sucked up from flowers, swallowed and carried back to the nest by foraging bees. Foragers of social bee colonies deliver the nectar to house bees that ingest it. House bees may use some for their own nutritional needs and also spread nectar around the colony by regurgitating it and offering it to other workers. Foragers receive food from nestmates just before taking off on a foraging flight, and workers feed larvae by regurgitating food into a cell. This has been described as the "**communal stomach**" of the social insect colony.

Most animals	Highly social insects
Cells	Cells
Organs	Organs
Organisms	Organisms
----	**COLONY or SUPER-ORGANISM**
Population	Population
Species	Species

3.6 Age-related behavioural development

A remarkable feature of insect social behaviour is the cooperation between individuals in a colony. Worker bees share the large number of tasks needed to maintain the nest. One of the mechanisms they use to divide up these tasks is age-related behavioural development (technically called temporal or **age polyethism**), in which workers naturally progress from one task to the next as they age (Figure 3-18).

Adults start as house bees, eventually graduating to being foragers when they are older. At first, they remain on the brood from which they emerged, building brood cells, feeding the young, handling food and making beeswax. They then progress to jobs that take them away from the core of the nest to guard the entrance or to circulate air by fanning their wings. Only towards the end of their life do they leave the hive and become foragers for nectar, pollen, water and propolis. This strategy makes sense because the riskiest jobs are given to older workers who are close to death anyway. On this point, I need to publicly declare that I am not recommending that human beings follow this strategy!

The progression of an individual bee through these stages is plastic, in the sense that they can remain doing one task for longer, or accelerate into the next task earlier, depending on the needs of the colony. They are even capable of returning to a previous task if the need is great.

The switch from hive work to foraging is a particularly dramatic change. The hive is dark, humid, safe, and maintained at a constant temperature. Foraging outside the hive exposes the bee to light, rain, predation, and wide temperature fluctuations. The bee is booted from a 24-hour cycle of continuous but gentle work to a strong circadian rhythm of highly energetic work during the daylight hours and rest at night. After a relatively short time foraging, most bees die in the field after failing to make it home.

We regularly refer to house bees and foragers in this book. Next time they are mentioned,

TABLE 3-2 Insect societies present an extra level of biological complexity, the super-organism.

BUILD BROOD COMB

➤ PROVISION BROOD

➤ HANDLE FOOD

➤ GUARD ENTRANCE

➤ REMOVE WASTE

➤ FORAGE

EGG
LARVA
PUPA

WORKER BEE PROGRESSES THROUGH A SEQUENCE OF TASKS AS SHE AGES...

BEE EMERGES
FROM CELL

BEE
DIES

HOUSE BEE — FIELD BEE

FIGURE 3-18 As an individual worker bee ages, it passes through a series of behavioural stages in which it first works in the nest and then progresses to tasks outside the nest. ARTWORK GLENBO CRAIG

spare a thought for the astonishing hormonal, neurological and behavioural changes that each worker must experience to allow her to play both roles during her life.

In this chapter, we have explored the social biology of two highly social insect groups, the honey bees and the stingless bees. These two groups have much in common: for example,

inflexible castes and perennial colonies founded by groups not individuals. But a close examination reveals many significant differences in social behaviour between honey bees and stingless bees (Table 3-3). (Differences in nesting behaviour are discussed in Chapter 4.)

TABLE 3-3 Summary of differences in social behaviour between honey bees and stingless bees.

Feature	Honey bees	Stingless bees
Reserve of virgin queens	No	Yes
Queen's mating frequency	Many	Once
Sub-families	Yes	No
New replacement queen	Selected by fighting of rival queens	Selected by workers
Laying of eggs by workers	Yes	In some species
Developmental period of the three castes	Different duration for each caste	Similar duration for all castes
Relative size of worker and drone cells	Unequal (drones are larger)	Equal

QUEEN
FEATURE

Images of the queen of *Tetragonula carbonaria* by **JAMES DOREY**

FIGURE 3-19 *(ABOVE)*

The queen on the brood comb surrounded by a "court" of attending workers. Note her tattered wing tips from a life of frequent wing signalling.

FIGURE 3-21

The queen signalling with her wings, a frequent behaviour important in communication with the workers.

FIGURE 3-20

The queen surrounded by a "court" of attending workers. Note the mature queen cell that has fallen from above and now rests on the young brood.

FIGURE 3-22

The queen. Note the collared cell ready for egg laying.

4 Nesting in the highly social bees

The highly social honey bees and stingless bees build complex nests for protection, to rear their young and to store their food.

Honey bees use their **hexagonal comb** to both rear young and store food, but stingless bees rear their young in **special brood cells** and store food in **large pots**.

Honey bees build their nests principally of **wax**, but stingless bees mix wax with plant resins to form **propolis**, their main building material.

Stingless bees protect themselves by constructing a strong nest wall and entrance tube, and with guard bees that bite and daub resin on intruders.

Honey bees maintain **tight control of temperature** in the nest. Stingless bees are less capable, but still exert moderately good control.

Honey bees **feed their young** regularly, but stingless bees mass provision the brood cells.

Honey bees re-use their **brood cells**, while stingless bees destroy and rebuild the cells for each use.

Honey bees **found a new colony** by the sudden swarming of many workers and the old queen. Stingless bees first build a new nest and gradually move in with a new queen.

The nests of honey bees and stingless bees differ in many fundamental ways. In this chapter, we enter the nests of the highly social bees and compare and contrast them.

But, before we do, let's define some terms.

A COLONY refers to a unit of bees cooperating to rear young.

A NEST refers to the physical structures, building materials and stored food of a bee colony in a natural location.

A HIVE refers to a colony of bees, their stored food and building materials, and the artificial structure built by humans to house them. The word "hive" is also commonly used to refer to the artificial structure without the bees but, in this book, I use the expression **HIVE BOX** for that.

Brood comb of the stingless bee
Tetragonula carbonaria. IMAGE **GLENBO CRAIG**

In nature, both honey bees [2] and stingless bees typically nest in cavities, usually in tree trunks, but also in rocky cliffs and artificial hollows. Nests of stingless bees are also occasionally found in ground cavities formed by the rotting of trees or the burrowing of other animals.

Stingless bees have little ability themselves to dig a nest cavity in the ground. Some species in the Americas and Asia build their nests exclusively in the nests of ants or termites, and these bees do excavate their nests, but this behaviour has not been seen in any Australian species.

[2] Please note, when comparing honey bees and stingless bees in this chapter and the previous one, I refer mainly to the European honey bee *Apis mellifera;* some of the details may be different for the Asian species of *Apis*.

4.1 Nest architecture

4.1.1 MAJOR COMPONENTS OF THE NEST

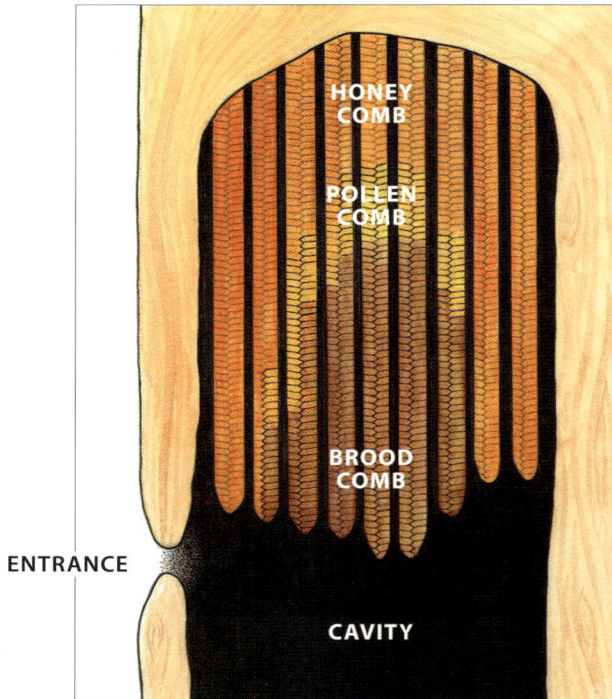

FIGURE 4-1 Typical honey bee nest in a tree cavity.
ARTWORK **GINA CRANSON**

Honey bees build just one type of **comb**, consisting of interlocking hexagonal **cells** (Figure 4-1, Figure 4-2). They build this comb in vertical sheets suspended from the top of their nest cavity. The hexagonal cells pack together efficiently and open to one side. This comb is used for all purposes: brood are safely reared in the cells towards the centre of the layers of comb, while food is stored around the outside. Pollen is stored close to the brood where it is used almost immediately to feed the young. Honey is stored in the outermost combs where it may be kept for months and so will not hinder the production of young if more cells are needed for brood.

FIGURE 4-2
A honey bee nest within a cavity, viewed from below and showing the layers of vertical comb and hexagonal cells.

IMAGE **ERIC TOURNERET**

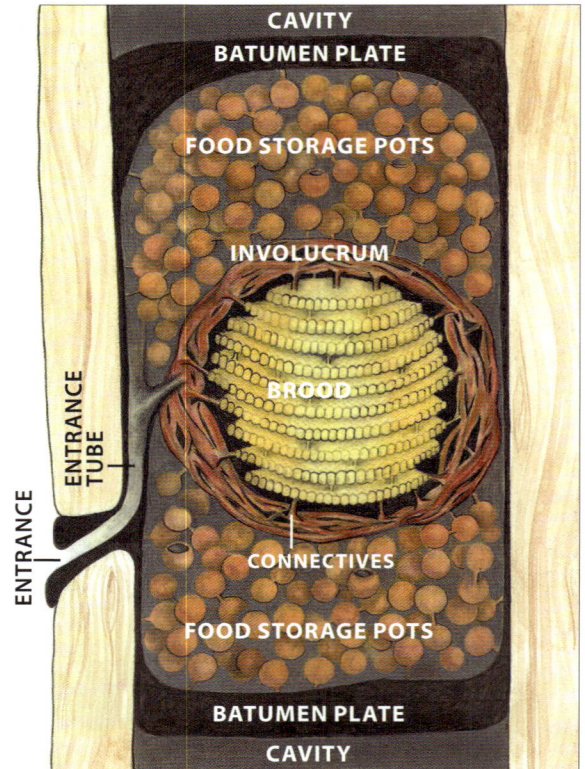

FIGURE 4-3 Typical stingless bee nest in a tree cavity. ARTWORK **GINA CRANSON**

In contrast, stingless bees build a distinct comb for rearing their young and a different kind for storing their food (Figure 4-3, Figure 4-4). The **brood comb** of stingless bees consists of many cells built in contact with each other. The **brood** cells can be arranged in various ways depending on the species, the most common arrangements being irregular **cluster** comb, horizontal layers of **regular comb** or the even, horizontal layers of **semi-comb**. The **brood** cells normally open at the top but, in some species, can open either at the top, the side or angles in between (Figure 7-9). The brood is normally grouped together in a distinct area of the nest called the **brood chamber**. This brood chamber is often surrounded by an **involucrum** which protects and insulates the brood (Figure 4-5).

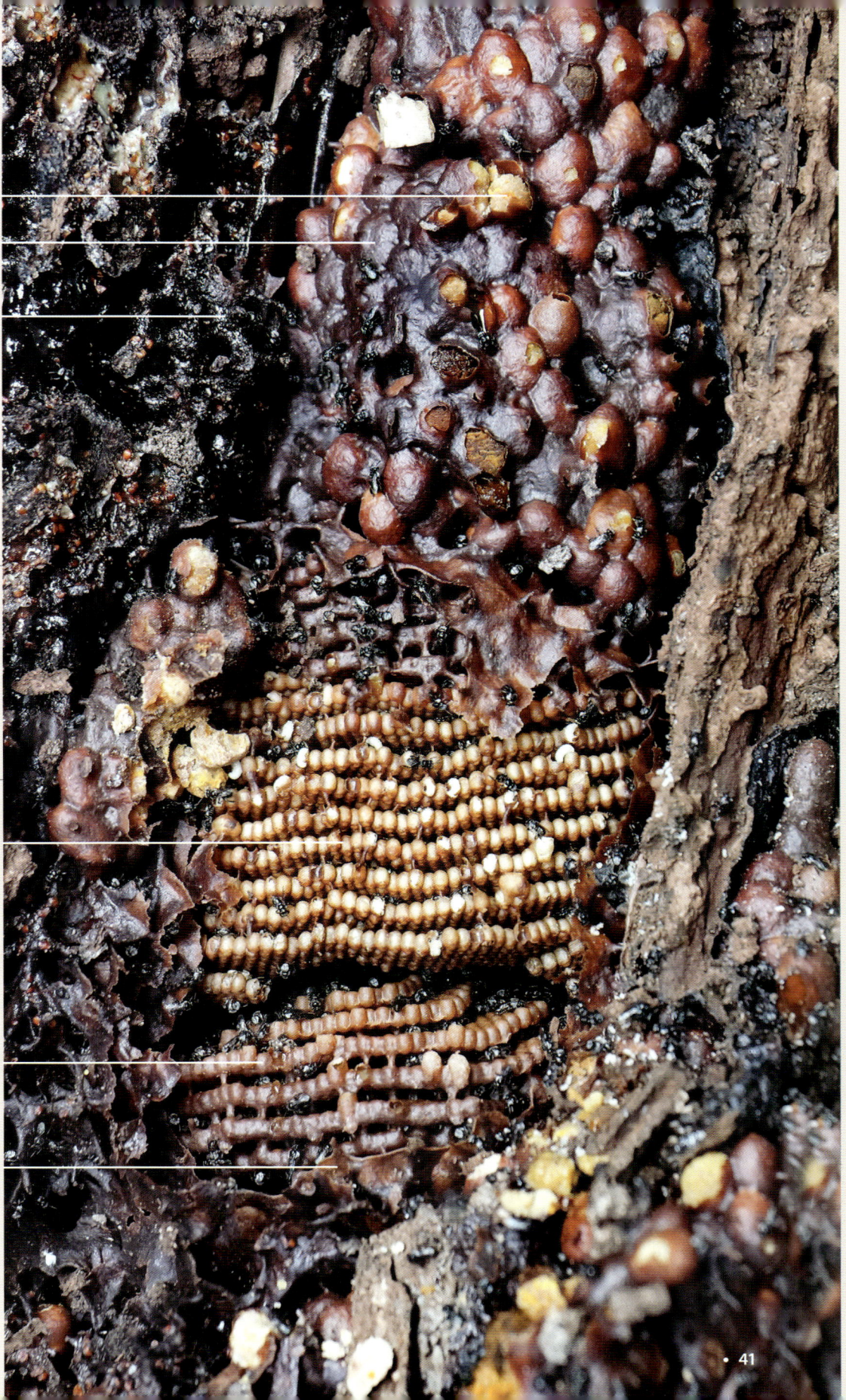

POLLEN POTS

HONEY POTS

BATUMEN PLATE

FIGURE 4-4
A stingless bee
(*Tetragonula
carbonaria*)
nest in an
opened log.
IMAGE
DAN COUGHLAN

MATURE BROOD

YOUNG BROOD

INVOLUCRUM

The food storage **pots** of stingless bees, which contain either honey or pollen, are not mixed with the brood but are separated and surround the brood. The food storage pots are conspicuously bigger than the brood cells. Typically the food pots are 10 mm – 15 mm in diameter, compared to the brood cells which are only 3 mm – 4 mm in diameter. The entire nest is surrounded by a **nest wall** (thickened to a **batumen plate** at the top and bottom). The major nest components are joined to each other by columnar structures called **connectives** (Figure 4-6). The entrances of honey bee nests are rarely modified but left open (Figure 4-1). Stingless bees reduce the size of the entrance and continue it as an internal tube that is highly defendable (Figure 4-3).

FIGURE 4-5
The multi-layered involucrum surrounding the brood of a *Tetragonula carbonaria* stingless bee nest.

FIGURE 4-6
The connective structures that stingless bees build to join the nest parts to each other and to the walls of the cavity.

IMAGE
GLENBO CRAIG

4.1.2 BUILDING MATERIALS: WAX, RESIN AND PROPOLIS

Stingless bees normally use propolis (a mix of beeswax and plant resins) as their building material (Figure 4-7). This mix is called "cerumen" in some stingless bee literature but, in this book, I use "propolis", the older established synonym. The ancient Greek word "propolis" is particularly apt for stingless bees because it means "before the city", and stingless bees use propolis to protect the nest entrance.

Beeswax is produced by glands on the abdomens of bees (Figure 1-4). Generally, the wax of stingless bees is a white to yellow substance, softer than the beeswax of honey bees, and with a lower melting point. But wax produced by different bee species varies; for example, *Tetragonula carbonaria* and *Tetragonula hockingsi* bees produce a darker wax than *Austroplebeia australis*, which produces a pale creamy or white wax.

Resin, a substance produced by plants, is harvested by forager bees (Figure 4-8). It is curious indeed how stingless bees can handle this sticky secretion. They collect it with their mandibles and front legs and transfer it to their hind legs for transport back to the nest (See resin-laden bee on page viii). In contrast, other insects, including ants, which are their close relatives, get stuck in resin. Stingless bees may be seen collecting artificial sticky materials such as perished rubber, sticky tape and wet paint. These materials have a consistency similar to natural resin and so are collected by stingless bees.

Plant resins are biologically active, containing two groups of chemicals of particular interest: **flavonoids** and **terpenoids**. Flavonoids are known to be antimicrobial and help to suppress the growth of nest pathogens that would otherwise flourish on the rich food reserves in the warm and humid conditions of the nest. Terpenoids are known to repel ants and other enemies.

Resin is sticky when first collected but sets hard with age. Wax is easily moulded, soft and retains its softness for years. The dark substance formed by mixing these two materials, propolis, has excellent properties as a building material: it hardens with age and is waterproof. It even inhibits the growth of microorganisms and deters enemies from entering the nest.

Honey bees build their nests primarily of pure wax. Honey bees in hives use propolis in only a limited way, although they do construct a propolis envelope on rough cavity walls when nesting in tree hollows. Stingless bees use a wax and resin mix for most nest structures. This is why the inside of a stingless bee nest looks dark and feels sticky, while a honey bee nest appears lighter in colour. The addition of resin to wax renders the stingless bee hives resistant to the ravages of the **wax moth**, a scourge of honey bee hives.

Stingless bees vary the ratio of the two components of propolis according to the needs of the structure. Their nest entrance is built of a **hard propolis** rich in resin so that it will set hard and also because resin contains terpenoids to repel enemies such as ants. Their **nest wall** (or **batumen**) is also high in resin because this structure needs to set hard and be strong. Brood cells and involucrum are made of a **soft propolis** rich in wax because they are constantly being rebuilt and need to be remoulded continuously. Including resin in brood cells can help to inhibit brood pathogens.

FIGURE 4-7 Worker stingless bees applying propolis to build connectives. IMAGE **NADINE ANDERSEN**

FIGURE 4-8 A tail end of a returning foraging bee with resin collected from plants, carried in pollen baskets. IMAGE **GIORGIO VENTURIERI**

FIGURE 4-9 The batumen plate, a thickened bottom nest wall of a *Tetragonula carbonaria* nest in a log.

4.1.3 DEFENSIVE STRUCTURES

The nests of socials bees offer rich, concentrated stores of proteins and carbohydrates, which are extremely attractive to a range of natural enemies. Not surprisingly, bees have developed a number of defences against these enemies. Honey bees rely largely on their stings. Stingless bees use a variety of alternative defensive mechanisms in the nest wall and around the nest entrance.

A strong wall (batumen)

Honey bees found new colonies in spaces that provide protection, but otherwise they do not modify those nest spaces. In contrast, stingless bees invest heavily in building strong physical barriers around their nests.

When first founding a new nest, stingless bees build an outer nest wall, the batumen. This nest wall defines the volume of the nest and separates it from the larger volume of the occupying cavity. It is constructed largely of propolis but also incorporates other materials such as pollen, faeces, wood and soil. It prevents the entry of natural enemies and insulates the colony from environmental stressors.

The nest wall is always a work in progress, added to and hardening over time. In very old nests, the wall can be several centimetres thick and very hard. The wall thickens mainly at the ends of the nest where it forms a partition, the batumen plate (Figure 4-3, Figure 4-9). Where the nest is in direct contact with the internal walls of the cavity, it is much thinner and is called batumen

lining (Figure 4-10). When the wood surrounding a nest decays and disintegrates, sometimes the nest wall is all that remains between the nest and the outside world.

The nest wall may have perforations into the hollow of the nest cavity that allow ventilation (Figure 4-11). In a well designed and built artificial hive box of an appropriate volume, the bees have no need to build a nest wall and so do not do so. But they may coat the inside of the hive box with a thin layer of propolis.

FIGURE 4-10
Batumen lining of the internal wall of a stingless bee hive. IMAGE **GLENBO CRAIG**

FIGURE 4-11
The batumen plate, the thickened external nest wall, of an *Austroplebeia essingtoni* nest, walling off the end of the nest within its hollow log home. Note the perforations to allow ventilation.

A secure nest entrance

Another key defensive structure is the entrance. Honey bees rarely modify the existing entrance to the cavity that holds their nest (often an old branch hole), but stingless bees are extremely particular about their entrance and put immense effort into forming it to their requirements.

Stingless bees apply propolis to reduce the size of the entrance to a size that suits them (Figure 4-12). The size is variable, but typically about 5 mm – 10 mm in diameter. They extend the entrance as a tube internally (Figure 4-3, Figure 4-13) and sometimes externally (Figure 4-14). The small size of the entrance and the accompanying entrance tube allow the bees to monitor incoming traffic. Guard bees wait at the entrance and along the entrance tube to inspect any incoming organisms and react against those that are not welcome. (Read more about this in "How stingless bees defend themselves" below.)

Some species extend their nest entrances with sticky layers of fresh resin to entangle intruders (Figure 4-14). Some species extend the tube in response to predator attack, particularly ants. Some, particularly *Austroplebeia* species, protect the nest entrance with a resinous curtain that they put up each night and take down again the next morning (Figure 4-15).

If their colony grows, stingless bees can widen their nest entrance by removing propolis. But if the entrance is naturally narrow they have only a limited ability to enlarge it, as their mandibles are not strong enough to scrape and remove solid wood. Peter Davenport has observed a nest that died because its host tree grew over the nest entrance and cut off access to the outside world.

FIGURE 4-13 Internal entrance tube of a stingless bee colony *(Austroplebeia australis)* in an open hive box. This tube is an internal extension of the entrance hole. IMAGE **CLAUDIA RASCHE**

FIGURE 4-14 External entrance tube of an *Austroplebeia australis* nest. These bees extend the tube in response to predator attack. Note the lurking predatory green tree ants. IMAGE **SIMON ROBSON**

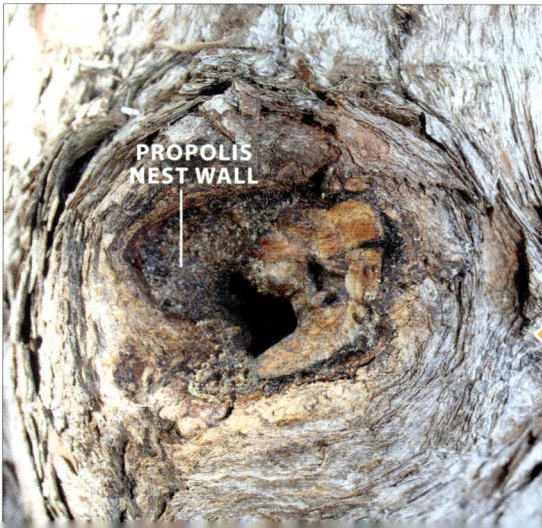

PROPOLIS
NEST WALL

FIGURE 4-12 Entrance hole of a stingless bee colony in a natural log, showing how they reduce the entrance to their preferred size using propolis.

FIGURE 4-15

The nest entrance of *Austroplebeia* species is closed by a resinous curtain each night.

IMAGE **GLENBO CRAIG**

4.1.4 ACCESS AND VENTILATION STRUCTURES

Stingless bees may build a **peripheral space** between the nest wall and other structures (Figure 4-16). This space is kept clear even when the rest of the nest is almost completely filled with comb and pots. It allows the bees to get from point A to point B with least obstruction. The peripheral space also serves a ventilation function, so it is a combined transport corridor and ventilation duct.

FIGURE 4-16

The peripheral space between the food storage pots and nest wall of a stingless bee colony in a hive box. This space extends all around the nest and connects to the internal entrance tube to allow movement of workers and ducting of fresh air.

IMAGE
GLENBO CRAIG

FIGURE 4-17
A typical arrangement of a *Tetragonula carbonaria* stingless bee nest, with the brood in the centre, yellow pollen pots at the front (*RIGHT*) and the amber honey pots at the back (*LEFT*)).

FIGURE 4-18 The food pots of *Austroplebeia* are made of wax, making them more translucent than those of other species. This translucency makes it easier to identify honey or pollen contents from the outside.

IMAGE **JEFF WILLMER**

4.1.5 FOOD STORAGE POTS

Excess food is stored in pots. Pollen and honey are not mixed but placed in separate pots. Stingless bees are capable of storing large reserves of food in their nests. I estimate that the approximately 5 kg of total weight of mature nests of *Tetragonula carbonaria* or *T. hockingsi* comprises roughly 2 kg of honey, 2 kg of pollen and 1 kg of building structures (propolis). The bees themselves, both immature in cells and mature adults, weigh very little. Honey pots are often concentrated at the back and top of the nest, whereas pollen pots are typically seen at the front of the nest (Figure 4-17).

Storage pots are built of propolis, with the ratio of resin and wax in the propolis varying between species. The species of *Tetragonula* use a resin-rich mix, while those of *Austroplebeia* store their food in pots that are almost pure wax (Figure 4-18).

The honey stored in the pots is the result of the ripening of nectar. This conversion involves the dehydration of the nectar and the addition of enzymes produced by the bees. A high sugar content and high acidity help to preserve this important energy source for later consumption. The honey is then sealed in pots. (Read more about honey in Chapter 12 "Honey from stingless bee hives".)

The large reserves of pollen kept by stingless bees (Figure 4-19) surprise many keepers of European honey bees. Honey bees typically keep only a few days' supply of pollen in their hives. But stingless bees are able to preserve

pollen for the long term. The tart taste of their pollen gives a hint as to how they achieve this. Microbiological assays of the pollen reveal a rich diversity of microbes. These "good microbes" ferment the sugar in the nectar the bees add to pollen, increasing its acidity and protecting it from spoilage by "bad microbes". This gives stingless bees the advantage of being able to collect lots of pollen when it is abundant and continue to use it to rear brood in high numbers all year round.

FIGURE 4-19 Numerous pollen pots in a *Tetragonula carbonaria* hive. IMAGE **MIKAYLA LAMBERT**

4.1.6 DEPOSITS OF RESIN AND WAX

In addition to food, stingless bees also store excess building materials. The resin collected from certain plants when abundant can be deposited for later use (Figure 4-20, Figure 4-21). Similarly, wax secreted by glands on the abdomen can be laid down in depots (Figure 4-22) for later mixing with the resin to form propolis.

FIGURE 4-20 This forager has collected resin of a red colour. IMAGE **EVA CACHIN**

4.2 How stingless bees defend themselves

Earlier in this chapter we discovered how stingless bees use defensive nest structures to help protect their colonies. The nest wall and entrance tube force natural enemies and bees from other nests to pass by the colony's guards to enter the nest. These guards can bite and apply sticky resin to unwelcome intruders. Generally, guards will bite soft-bodied intruders (Figure 4-23) but apply resin to hard-bodied ones (Figure 4-24). Resin has been shown to be effective against the well armoured small hive beetle, which causes greater losses to honey bee hives than to stingless bee hives.

FIGURE 4-23
Defending workers of *Tetragonula carbonaria* biting attacking workers from another colony.

IMAGE **PAUL CUNNINGHAM**

FIGURE 4-21 Deposits of resin (in this case, from the cadaghi tree, *Corymbia torelliana*) stored in the nest for later use. IMAGE **GLENBO CRAIG**

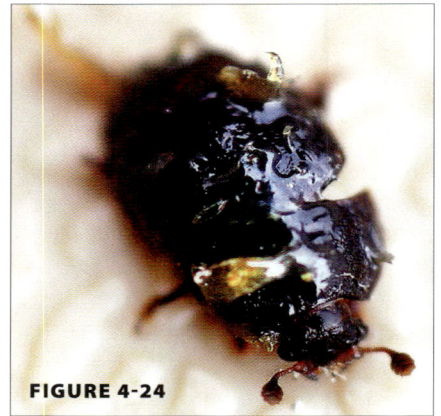

FIGURE 4-24

Resin applied to a small hive beetle that attempted to enter a hive of *Austroplebeia australis*. IMAGE **MEGAN HALCROFT**

FIGURE 4-22

A stingless bee adding wax secretions to a wax depot in the hive.

IMAGE
GIORGIO VENTURIERI

Stingless bees have minimal ability to defend themselves against humans and other large predators. Australian species only attack when their nest is opened, at which time they may defend themselves vigorously by biting softer parts of the body. (Read more about this in "Protecting yourself against the defences of stingless bees" in Chapter 13.)

Why would stingless bees have lost their sting? We know it happened a long time ago, as a fossil stingless bee in amber has been dated to about 70 million years ago. Perhaps we can find a clue in the closely related ant family. Loss of stings is very common in ants. In fact, stings are mostly retained in ant species that use them to attack their prey and mostly lost in ants that could use them for defence. It appears that a sting can be more of a liability than an advantage, perhaps because of the high cost of producing it. This may be why stingless bees evolved as they have.

Another form of "defence" is keeping a hygienic and healthy hive. In stingless bee colonies, worker bees are vigilant against potential infections in larvae and quickly remove any brood that die. Jenny Shanks has shown just how quickly *Tetragonula carbonaria* can perform hygienic behaviours. She artificially damaged larvae (killing via a pin prick) and watched how quickly the workers detected and removed the dead brood (Figure 4-25). *Tetragonula carbonaria* workers have superior hygienic behaviours, as they detect dead larvae faster than many other stingless bee species. This results in cleaner hives with a low probability of contracting brood infections.

Latrines

In what may at first appear to be contrary to the above-mentioned hygienic behaviour, stingless bees defecate and deposit wastes and corpses in the nest in specific areas. Strangely, these latrines are sometimes found beneath and close to the brood chamber. This may provide a way for beneficial microbes to transfer to newly emerged adults.

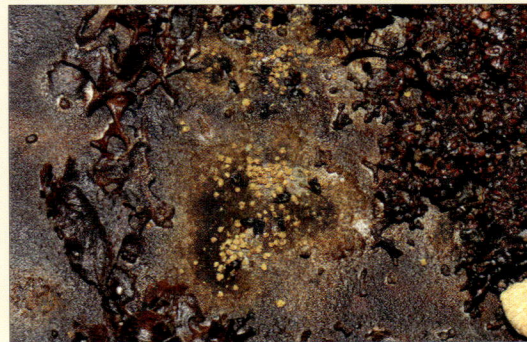

TOP IMAGE **JAMES DOREY**
LOWER IMAGE **GIORGIO VENTURIERI**

FIGURE 4-25 Marked and pin-killed brood cells *(LEFT)*. Adult pulling diseased larva from cell *(MIDDLE)*. Complete removal of killed pupae and their cells *(RIGHT)*.

IMAGES **JENNY SHANKS**

4.3 Temperature regulation in social bee nests

Both stingless bees and honey bees attempt to maintain nest temperature within a favourable range. Honey bees regulate the brood temperature within the range 32–36°C. Stingless bees are less capable, with temperatures recorded in the brood of *Tetragonula carbonaria* ranging from 20°C to 40°C, and from 0°C to 38°C in *Austroplebeia australis*.

Temperature regulation is one of the ways that a colony of insects resembles a more complex

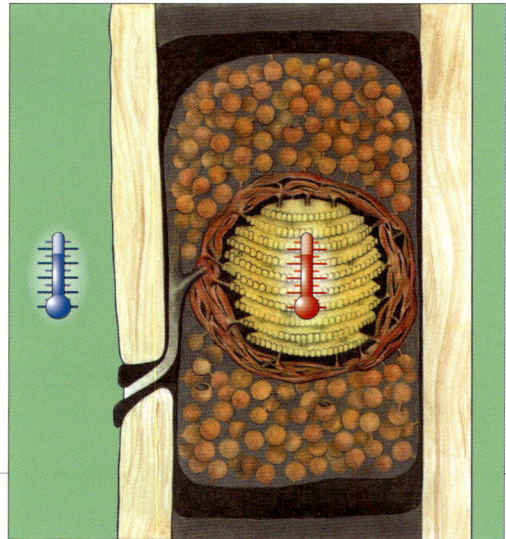

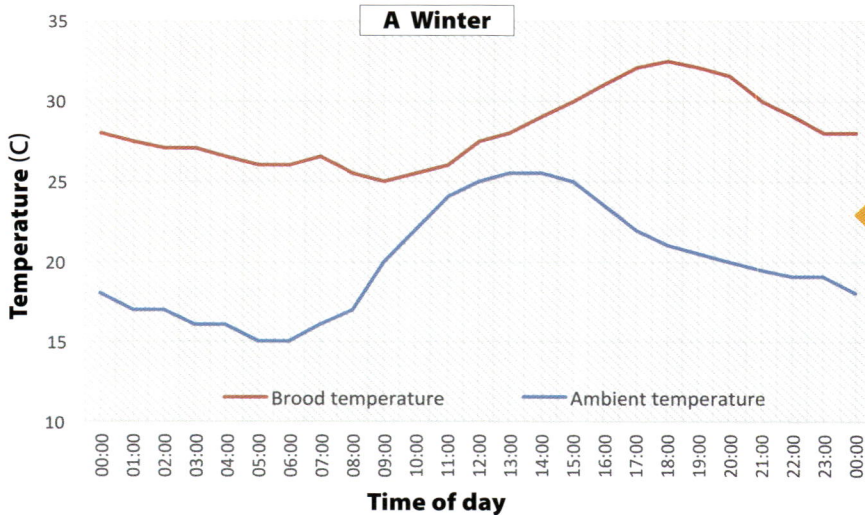

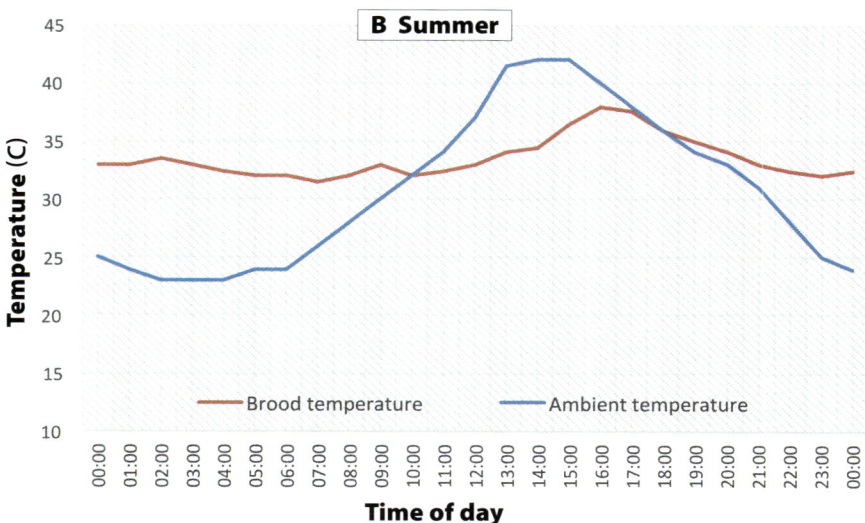

FIGURE 4-26

Typical 24-hour cycle of the brood temperature of a colony of the stingless bee *Tetragonula carbonaria* (red line), compared with the temperature outside the nest (blue line).

A – Winter.
B – Summer.

(Compiled from the data of K. Amano, B. Luttrell, T. Heard, G. Venturieri and J. Klumpp)

ARTWORK
GINA CRANSON
AND **GLENBO CRAIG**

individual organism. Mammals, being warm-blooded, are able to maintain a constant body temperature by metabolic warming and, in some cases, evaporative cooling (e.g. sweating). Social insects also do this by a collective activity called **social thermoregulation**. This allows bees to rear their brood at a constant rate even when the weather outside is extreme.

How do they do this? Well, when the ambient temperate is low, bees generate heat by their metabolic activity. If the nest is well insulated, the heat is trapped and keeps the colony warm. Honey bees have an additional strategy in extremely cold climates of ceasing brood rearing, congregating close together, and waiting out the winter. Stingless bees do not have this ability to hibernate. But, in cool conditions, they retain the heat generated by their metabolic activity, and extra workers will congregate on the brood to keep the temperature there higher than ambient (Figure 4-26 A). This is aided by the involucrum, which is marvellously structured for insulation.

Species of stingless bees that do not build a thick, multi-layered involucrum do not have the same ability to maintain high brood temperatures. Megan Halcroft has shown that the brood of *A. australis*, which only builds a single-layered, partial involucrum, experiences dramatic changes in temperature depending on external conditions. In this species, the brood temperature can fall to close to 0°C and yet, remarkably, the bees survive.

In hot conditions, social bees attempt to keep their nest cool. Again, honey bees manage this effectively by employing **evaporative cooling**. They collect water from their environment, disperse it around the nest and fan their wings to cool things down. Stingless bees are not capable of evaporative cooling. But good insulation and a place in the shade will protect a colony from ambient extremes of heat. Figure 4-26B shows how conditions in the brood chamber inside a hive are buffered from external extremes. (Read more about managing the temperature in stingless bee hives in "Protecting colonies from weather extremes" in Chapter 13.)

Stingless bees *can* modify nest temperatures by active **ventilation** at the entrances. Worker bees fan their wings while facing outwards towards the entrance to draw air into the nest. In particular, they can draw in cool air if the colony is overheating. **Air circulation** is also necessary for gas exchange. The intense concentration of bees in a space with only one or a few openings can result in a depletion of oxygen and the dangerous build-up of carbon dioxide.

In addition to temperature, social bees exert some control over the **humidity** in the nest. Typically, the air in the nest is humid. When the bees are actively making honey, it becomes particularly humid. Much of this humidity is removed in the process of ventilating the nest. But some may condense inside the hive. This is ingested by workers, which then move to the nest exits and spit out the excess water. As a result, you may see brown stains dripping down from the entrance or the ventilation hole of a hive (Figure 4-27). The dark colour comes from the water-soluble fractions of propolis.

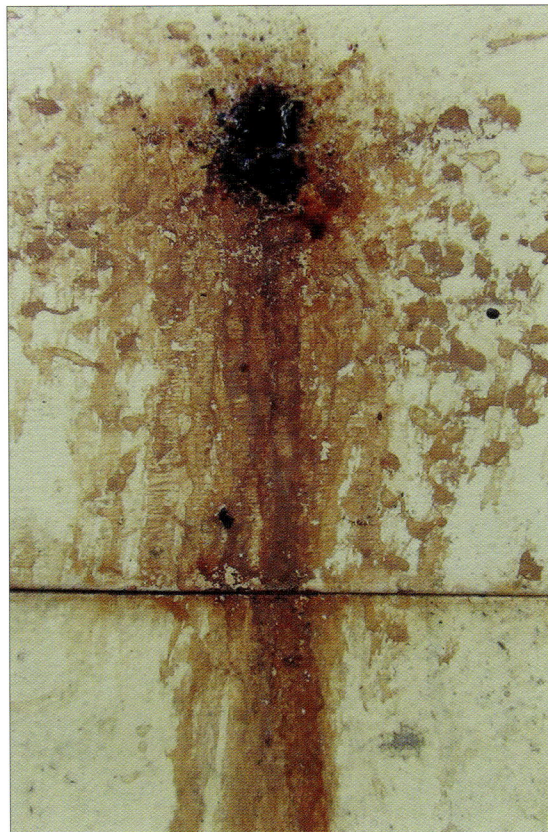

FIGURE 4-27 Brown stains below the ventilation hole of a hive, the result of removal of excess water.

4.4 Brood rearing in social bees

Social bees rear their young in individual cells in the central area of the nest, which offers the best insulation and protection (Figure 4-1, Figure 4-3). Here, each individual bee develops from egg to adult, a process known as **ontogenesis**.

Let's explore this process for honey bees and stingless bees, and see just how different their strategies for "child-rearing" are.

4.4.1 BROOD REARING BY HONEY BEES

In honey bees, the process begins when the queen lays her egg in an existing cell (Figure 4-28).

Developmental times and feeding differ between castes, with the following details applying to a worker bee.

Honey bee development

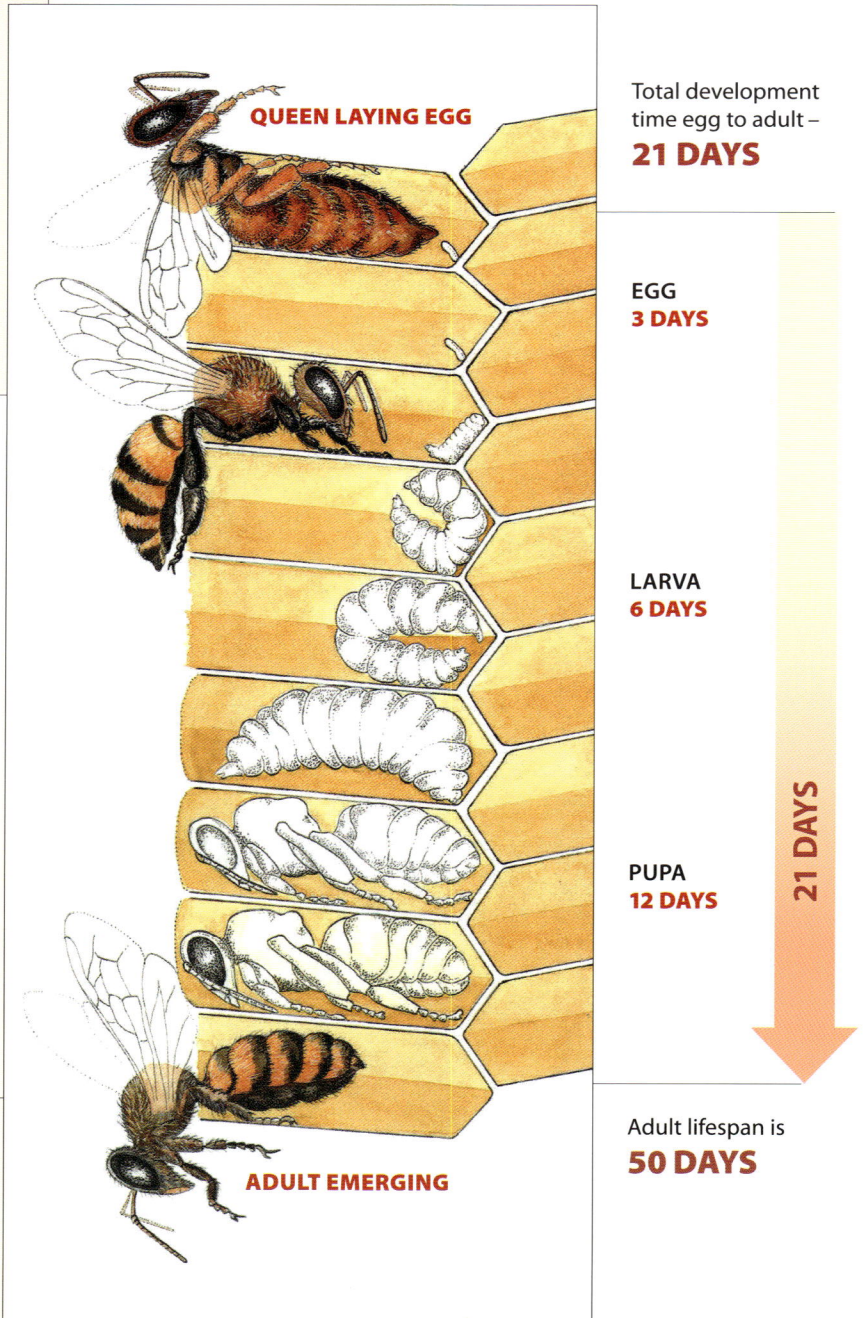

QUEEN LAYING EGG

Total development time egg to adult –
21 DAYS

**EGG
3 DAYS**

**LARVA
6 DAYS**

**PUPA
12 DAYS**

21 DAYS

ADULT EMERGING

Adult lifespan is
50 DAYS

FIGURE 4-28
Cross-section through comb of honey bee showing the process of rearing brood.
ARTWORK **GINA CRANSON**

After three days, the egg hatches and the larva is ready to feed. Workers feed the larva progressively by regurgitating one meal at a time. This **progressive provisioning** of larvae is not unique to honey bees, but is practised only by a few groups of bees.

The food is a glandular secretion (**royal jelly**) for the first three days and then a mix of pollen and honey for the remaining three days. The cell is then closed (capped) by the workers. The larva stops feeding, spins a silk **cocoon**, converts to a **pre-pupa**, defecates and pupates. The pupal stage terminates with the emergence of the adult. The total development time from egg to adult is 21 days for workers. Ontogenetic development times for social insects are very consistent because the temperature is held within a narrow range in the nest. The larvae destined to become queens receive royal jelly for their entire larval feeding period and are reared in specially constructed cells.

Following the emergence of an adult honey bee from its natal cell, the cell is cleaned out and re-used. The queen accepts this cell for reuse and lays an egg inside (Figure 4-29).

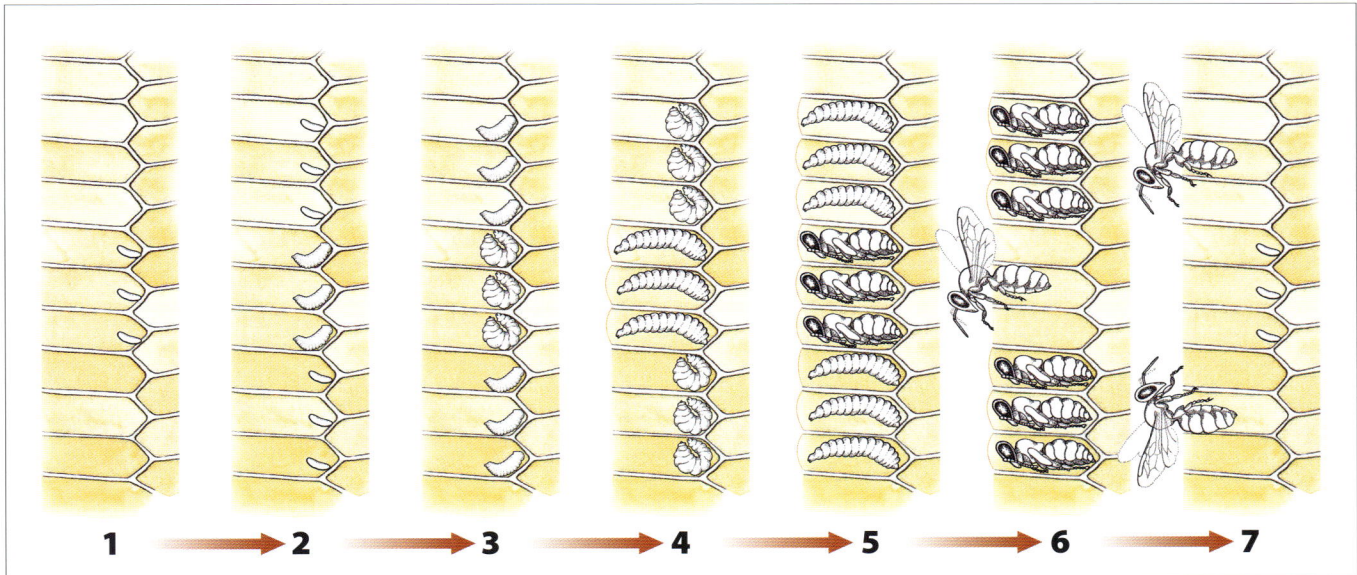

FIGURE 4-29 Cross-section of brood comb of honey bees showing a sequence of developing stages and the final reuse of cells. ARTWORK **GINA CRANSON**

Honeybee brood, showing both capped cells, and advanced larva being fed. Shiny cells on the left hold stored pollen.
IMAGE
GLENBO CRAIG

Stingless bee development

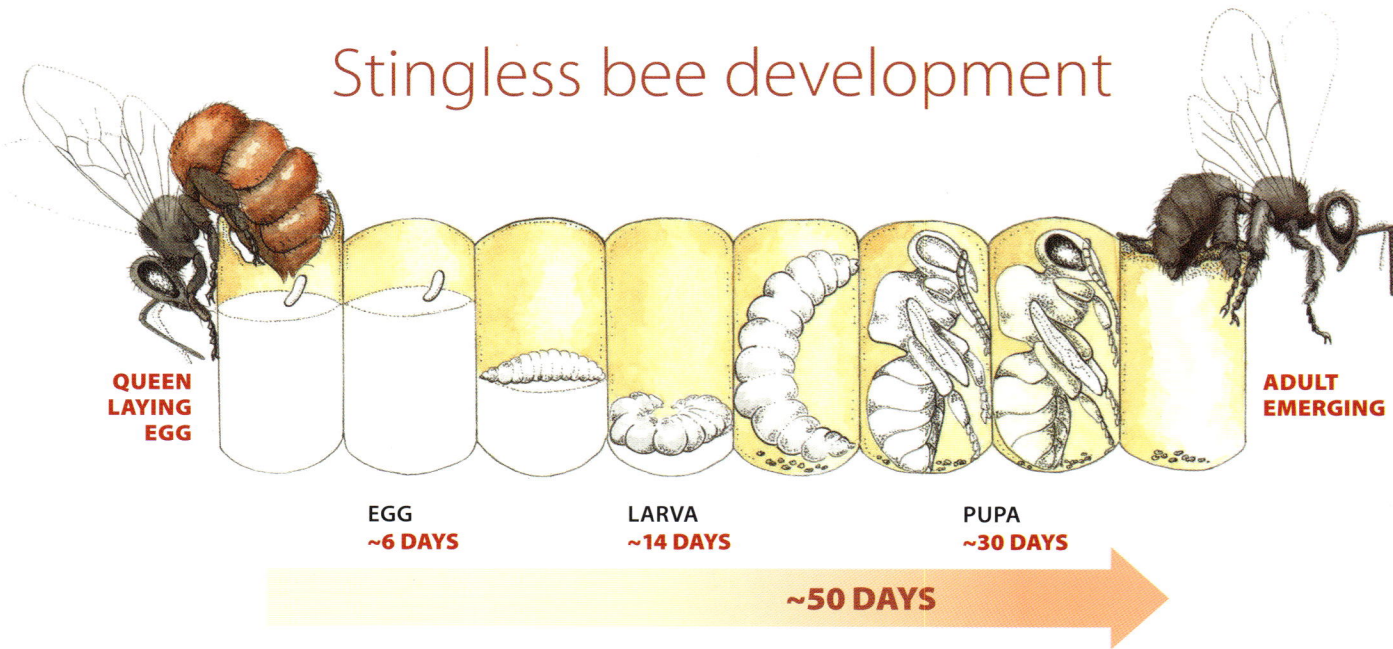

QUEEN
LAYING
EGG

EGG
~6 DAYS

LARVA
~14 DAYS

PUPA
~30 DAYS

ADULT
EMERGING

~50 DAYS

Total development
time egg to adult –
~50 DAYS

FIGURE 4-30 Cross-section through comb of stingless bee showing the process of rearing brood. ARTWORK **GINA CRANSON**

Adult lifespan is
~100 DAYS

4.4.2 BROOD REARING BY STINGLESS BEES

Brood provisioning and development

Stingless bees rear their young differently. A cell is **mass provisioned** by nurse workers regurgitating food to about two-thirds of the capacity of the cell, which is sufficient to feed the larva for its entire development to a pupa (Figure 4-30, Figure 4-31). This food is a semi-liquid mix of pollen, honey and glandular secretions (Figure 4-32). The nutritious glandular secretion is called **brood food** and is the equivalent of the royal jelly of honey bees, but all bees, including workers and males, receive this food.

The queen lays an egg on the provisions. The cell is then immediately capped so that the larva can develop in a closed cell. This resembles the ancestral nesting behaviour of the solitary bees, which also cap cells immediately after provisioning and laying an egg, and not the highly derived behaviour of modern honey bees. (Read more about this in "What are bees?" in Chapter 1.)

FIGURE 4-31 Nurse bees with their heads in newly constructed brood cells, inspecting and preparing to provision them, prior to the queen laying the egg.

IMAGE **GLENBO CRAIG**

FIGURE 4-32 Brood cells of the stingless bee *Tetragonula carbonaria*. Three open cells are under construction. Three cells have been opened to view the provisioned food with egg laid on top.

FIGURE 4-33 Cross-section of the brood comb of *Tetragonula carbonaria* stingless bees showing the cell contents with eggs at the top and C-shaped larvae in older cells at the bottom. IMAGE **JEFF WILLMER**

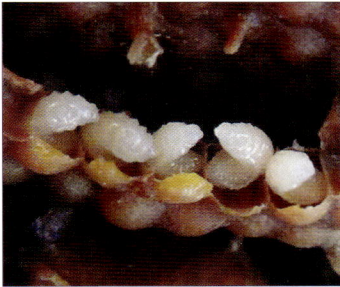

FIGURE 4-34 Brood cells opened to show immature bees at the final larval stage.

FIGURE 4-35 Cells of the oldest stage in the brood: the pupae. Workers have removed the soft propolis so that the cocoons are visible. The cocoons are semi-transparent, revealing the pupae inside. The pupae are positioned with their heads up as shown by the small eye spots on the top side. The faecal "meconium" stain is visible on the bottom side. IMAGE **MIKAYLA LAMBERT**

Like honey bee larvae, the larva (Figure 4-33, Figure 4-34) stops feeding, spins a silk cocoon, converts to a pre-pupa, defecates and pupates (Figure 4-35, Figure 4-36, Figure 4-37). After approximately 30 days, an adult emerges from the pupal cocoon.

Each of the immature stages of egg, larva and pupa is longer than for honey bees, with a total of approximately 50 days required to develop from egg to adult. For the first few days of their life, adults are pale and then darken to take on their full mature adult pigmentation. The pale adults are called **callows** and are often seen walking and working on the brood comb (Figure 4-36, 4-38).

FIGURE 4-36 The pupal brood with cocoons removed to show the individuals within. The darkening eyes are visible, along with other adult structures such as head and legs. The adult in the centre is a recently emerged callow that has not yet darkened.

FIGURE 4-37 The fragile cocoon removed and laid to the side to show the pupae inside.

FIGURE 4-38 A newly emerged adult (callow); she will become darker over the next few days.

IMAGE **GLENBO CRAIG**

Both stingless and honey bee larvae do not defecate until they have consumed the last of their food and spun their cocoon, so the faecal meconium is included in the cocoon (Figure 4-35). (In honey bees, the meconium is removed for cell reuse, but the cocoon remains in the cell.) Eventually, many layers of cocoons build up, giving the brood comb a dark appearance and reducing the internal volume of each cell. In stingless bees, the cocoon containing the meconium is removed and deposited outside the nest. (Read more about this in "Removal of waste garbage pellets" later.)

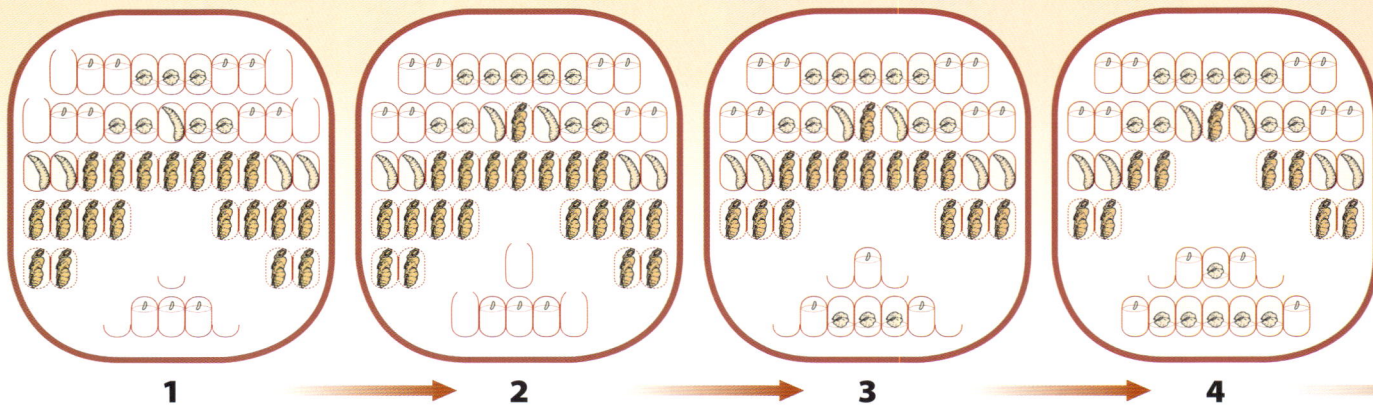

FIGURE 4-39 Eight sequential stages in the cycle of development of stingless bee brood comb. Note how cells are destroyed after each use and then rebuilt. ARTWORK **GINA CRANSON**

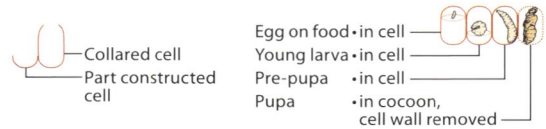

Collared cell
Part constructed cell

Egg on food • in cell
Young larva • in cell
Pre-pupa • in cell
Pupa • in cocoon, cell wall removed

Brood cell dynamics

Another key difference between the groups of highly social bees is **single use versus reuse of brood cells**.

Honey bees re-use their brood cells for the life of the nests (Figure 4-29). But the brood cells of stingless bees are **not** re-used. Here's how stingless bees do it. A new cell is constructed next to a recently completed one (Figure 4-39). The part of the brood where the new cells are being built and provisioned with food is called the **advancing front** (Figure 4-40). (Note that the newly completed open cells have higher walls than the adjacent cells. Cells at this stage are described as **collared**. The collar provides the material that will be folded down to form the cap of the cell when it is closed.) When complete, the cells are mass provisioned by workers. Then the queen immediately lays an egg on top of the food mass (Figure 4-32).

Over the next 50 days, the larva emerges from the egg, consumes its food, spins a silk cocoon and pupates inside the protective cocoon. Then a remarkable behaviour takes place: workers remove the soft propolis top and bottom of the cell and also the side walls when they are accessible (Figure 4-39, Figure 4-41). Only stingless bees and no other bee removes the cell wall in this way. The cocoon is exposed, creating a different appearance from the cells of younger brood (Figure 4-42, Figure 4-43, Figure 4-44).

FIGURE 4-40 A view looking onto the advancing front of *Tetragonula carbonaria* stingless bees. A batch of new cells has been constructed around each edge of the flat spiral comb. Note the collared cells. IMAGE **MIKAYLA LAMBERT**

FIGURE 4-41 *Left :* Brood cells at the larval stage showing the gradual removal of soft propolis from the older larval cells in the top left compared to the younger cells at the bottom right. *Right :* Cocoons appear silky rather than waxy, after removal of propolis. RIGHT IMAGE **JAMES DOREY**

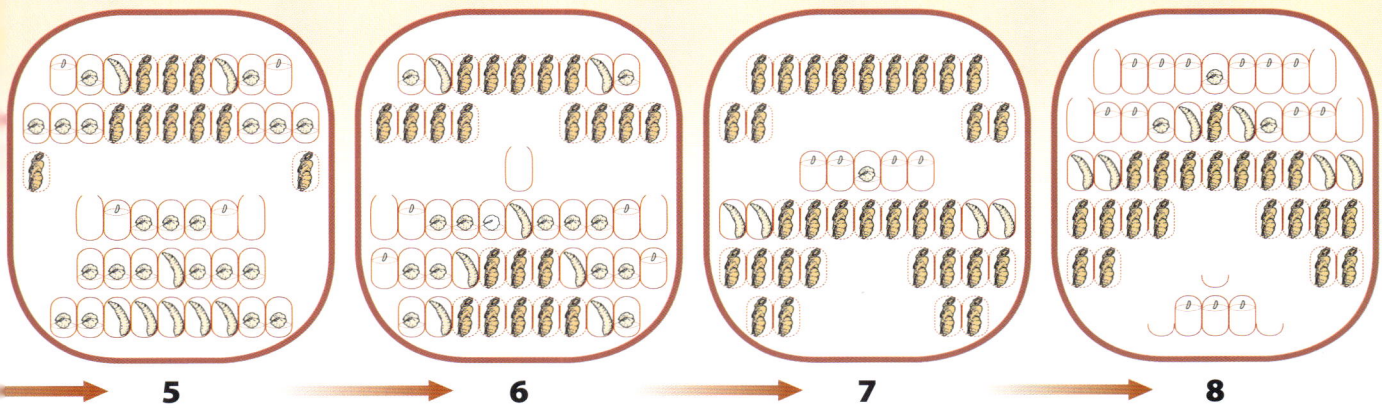

5 → **6** → **7** → **8** →

Stingless bees (at least those in the *Tetragonula* genus) pupate in their cell with their heads up. But, in a small proportion of cases, they pupate with their heads down. If you look carefully at mature brood, you may see some cells in which the meconium stain is above and the eye spots are below (Figure 4-43).

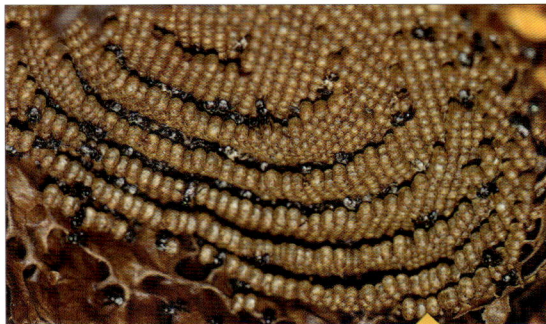

FIGURE 4-42 A side view of the brood chamber of *Tetragonula carbonaria* stingless bees. This shows the gap where the new cells are being built up from the bottom half of the brood (the advancing front) and how the oldest cells at the bottom of the top half of the brood are disappearing (the retreating edge) as the adults emerge.

RETREATING EDGE
ADVANCING FRONT
BROOD CELLS CONTAINING EGGS
BROOD CELLS CONTAINING LARVAE
BROOD CELLS CONTAINING PUPAE

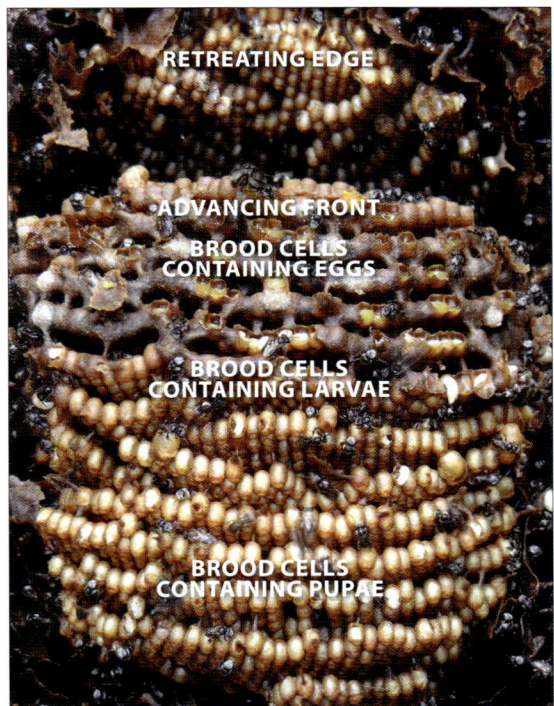

ADVANCING FRONT
BROOD CELLS CONTAINING EGGS
BROOD CELLS CONTAINING LARVAE
BROOD CELLS CONTAINING PUPAE

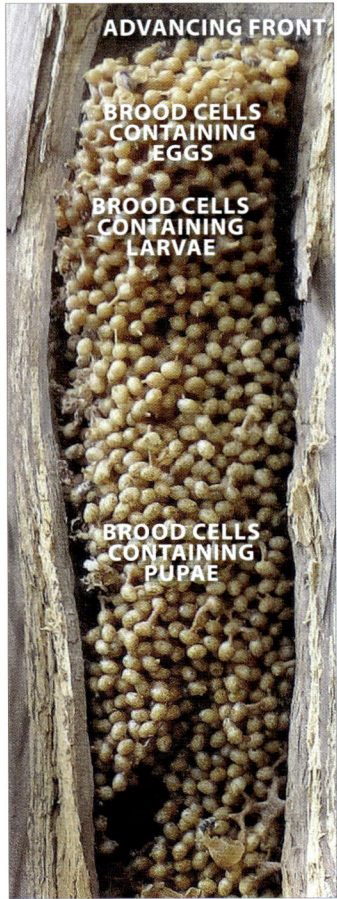

FIGURE 4-43
A view of the retreating edge of *Tetragonula carbonaria* stingless bees. The soft propolis has been removed from these cells and the adults are emerging.
IMAGE
NADINE ANDERSEN

FIGURE 4-44
The newly built brood cells of this *Austroplebeia essingtoni* nest look waxy, while the old brood cells, containing pupae, appear duller because the cell wall has been removed and only the cocoon is visible.

Multiple advancing fronts can be seen in stingless bee brood, particularly when the nest cavity is narrow. The narrow diameter creates a small area for building the advancing front and forces the nest builders to work at multiple fronts. I have observed a colony of *T. carbonaria* in a tree fern trunk that was 2 m long but only 60 mm in diameter.

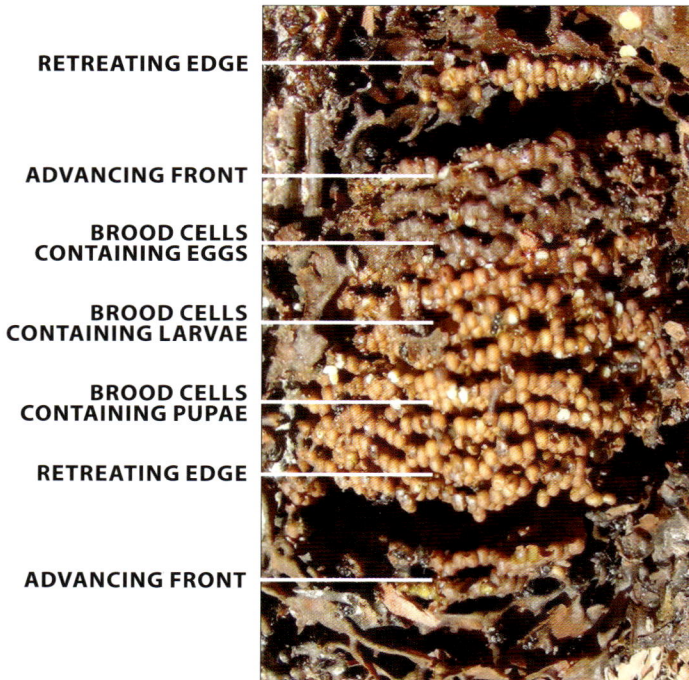

RETREATING EDGE

ADVANCING FRONT

BROOD CELLS CONTAINING EGGS

BROOD CELLS CONTAINING LARVAE

BROOD CELLS CONTAINING PUPAE

RETREATING EDGE

ADVANCING FRONT

The brood was 400 mm long with four advancing fronts. *Tetragonula hockingsi* may also display more than one advancing front (Figure 4-45). Multiple advancing fronts do not mean there are multiple queens. The single mated queen in stingless bee nests moves through the brood chamber visiting multiple advancing fronts.

Brood rearing in cluster brood builders

I have used *Tetragonula carbonaria* species as an example of brood rearing in stingless bees, but other species present different brood architectures. Instead of regular horizontal comb, most Australian and Asian species form their brood in cluster form.

All species of *Austroplebeia* (except *A. cincta*, which has concentric layers of brood cells) and some species of *Tetragonula* show a cluster arrangement of brood. The cluster arrangement shows little regularity in the position of the cells in relation to each other. They resemble a bunch of grapes, with the cells loosely packed and only touching each other at a few random contact points. But they still have an advancing front, the part of the brood where new cells are being built.

FIGURE 4-45 Side view of a *Tetragonula hockingsi* brood chamber, from near Townsville, in a log, showing two advancing fronts.

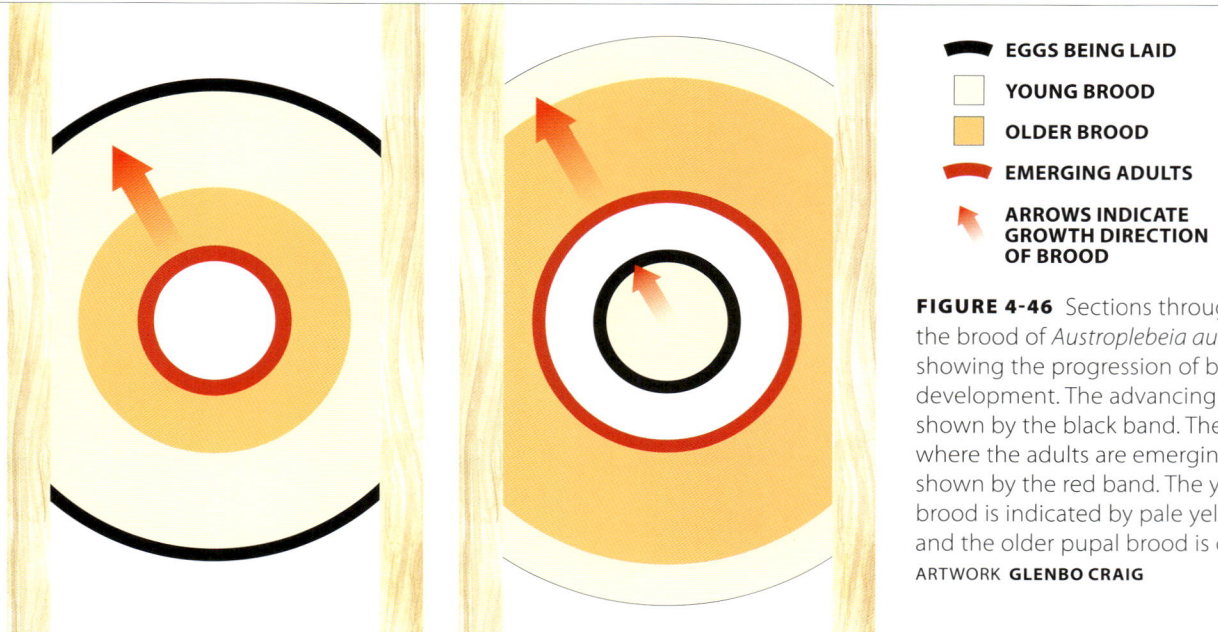

EGGS BEING LAID

YOUNG BROOD

OLDER BROOD

EMERGING ADULTS

ARROWS INDICATE GROWTH DIRECTION OF BROOD

FIGURE 4-46 Sections through the brood of *Austroplebeia australis*, showing the progression of brood development. The advancing front is shown by the black band. The area where the adults are emerging is shown by the red band. The younger brood is indicated by pale yellow, and the older pupal brood is orange.
ARTWORK **GLENBO CRAIG**

FIGURE 4-47

The advancing front of an *Austroplebeia cassiae* nest.

See how the brood cells can open downwards, the brood food is thick enough to stick to the base of the cell.

IMAGE **DAN COUGHLAN**

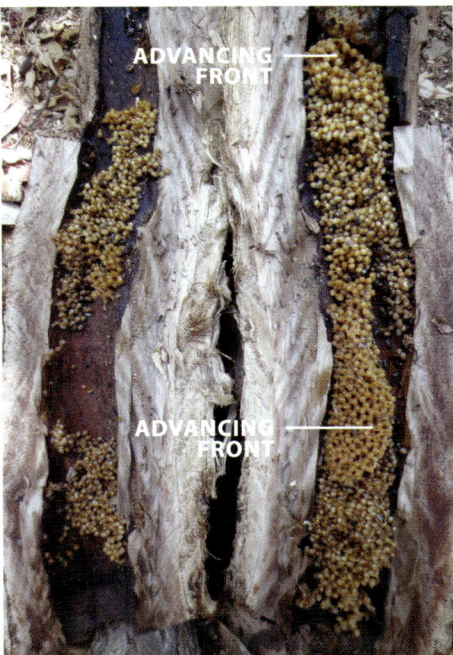

FIGURE 4-48 This nest of *Austroplebeia essingtoni* in an opened hollow log has two advancing fronts, one moving upwards and the other downwards.

The advancing front of the cluster builders defies gravity. It does not progress upwards but can move in any direction. For example, in *Austroplebeia*, the advancing front will often take the form of an expanding sphere (Figure 4-46, Figure 4-47). When this sphere reaches the outside of the brood chamber, it will start in the centre again where a space is starting to form because of the emergence of the oldest brood. This pattern may not be regular or symmetrical, with the advancing front forming only part of a sphere. In this case, the front may grow in a particular direction. That direction is not necessarily ever upwards, as it is in *T. carbonaria* and *T. hockingsi*, but can be downwards or sideways (Figure 4-48). When the colony has plenty of space, the brood may move its way around the nest, led by the advancing front. This process is assisted by the fact that the species that build cluster combs usually only have a partial involucrum (around the older brood) and so are not constrained to a fixed location.

Tetragonula hockingsi is highly productive as shown by the large number of open brood cells *(ABOVE)*.

Austroplebeia australis rears fewer cells per day (*TWO IMAGES TO THE LEFT*).

Brood productivity

Colonies of stingless bees rear brood all the year round. Unlike honey bees, they are not able to enter **hibernation** and suspend brood production.

However, colonies may vary the **rate of brood production** according to seasons and food availability. A colony of *Tetragonula carbonaria* will normally rear about 300 new adults per day. Hence a similar number of cells are constructed, provisioned and laid with an egg each day. Workers prepare cells synchronously in batches of around 70. Each batch takes about five hours, so approximately five batches are built per day. *Tetragonula hockingsi* are capable of even higher production but *Austroplebeia australis* colonies only rear about 50 per day. The queen, of course, has to be capable of laying this many eggs. In comparison, a honey bee queen can lay up to 1,500 eggs per day.

4.4.3 REMOVAL OF WASTE GARBAGE PELLETS

Rearing brood creates a waste product made up of the cocoons and meconium left by larvae. In addition, while adult honey bees leave the nest to defecate, stingless bees defecate in the nest. All of these wastes are bundled up and transported out of the nest in the form of a pellet.

Bees can often be seen removing garbage pellets from their nests. Workers that specialise as garbage removalists are commonly visible during periods of high activity, carrying an orange to brown garbage pellet in their mandibles to deposit outside the nest (Figure 4-49). They normally fly away with this pellet and dump it a distance from the nest.

However, when it is too cold to fly, bees may simply drop garbage pellets out of the nest entrance so that they pile up below (Figure 4-50). In bad weather when bees cannot go outside at all, these garbage pellets may accumulate inside the nest (Figure 4-51). When the weather improves, they then dump a lot of pellets in a short time. Ample waste removal is a good sign because it indicates that a colony is successfully rearing adults.

FIGURE 4-50 Garbage pellets removed from a hive and accumulating beneath the entrance.

FIGURE 4-51 Accumulation of garbage pellets inside the nest, near the entrance tube, of a colony of *Tetragonula hockingsi*.

FIGURE 4-49 A worker gripping a garbage pellet in her mandibles in the course of removing it from the nest.
IMAGE **TOBIAS SMITH**

FIGURE 4-52 A swarm of honey bees that has separated from the original colony and is attempting to found a colony in a new location. IMAGE **GLENBO CRAIG**

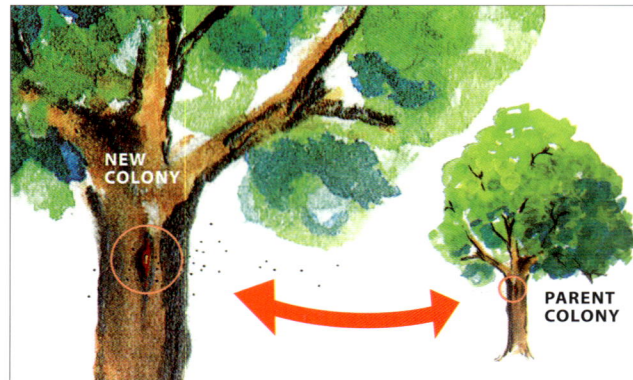

4.5 Founding a new colony

When a social bee colony is strong and environmental conditions are good, it may attempt to found a new colony. This is the natural way a colony reproduces. In honey bees, this process is also known as colony fission or swarming (Figure 4-52). (Note that "swarming" in stingless bees normally refers to another process; see "Fighting Swarms" in Chapter 13 for more information.)

In the case of honey bees, approximately half of a colony's bees prepare to leave the parent nest, then suddenly depart, taking the existing mated queen with them to found a new colony. The swarm keeps moving until it finds a suitable nest site. The parent nest becomes queenless but has prepared for this event by building queen cells. One of the emerging virgin queens kills the others, takes a mating flight and returns as the new mated queen. The split between original and new colonies is sudden, quick and total.

Stingless bee colonies employ a reproductive strategy that is different from honey bees in several significant ways.

When they perceive that they are strong enough to reproduce, they find a new nest site and make it suitable for a new colony. They build a nest wall to create an appropriate volume (often around 4–8 litres). They secure the entrance by reducing the natural size of the cavity entrance and building an internal entrance tube. They transfer propolis from the original nest as building and sealing material. They stock the new nest with provisions collected from the environment. Bees flow daily to and from the parent nest as individuals (Figure 4-53), not in a single swarm like honey bees.

This establishment phase may continue for weeks or months. It is not usually obvious that the parent colony is founding a new colony, but

FIGURE 4-53 When founding a new colony, the workers of stingless bees move from old to new location over time as individuals, not at once in a mass like honey bees. ARTWORK **GINA CRANSON**

it may be announced by workers leaving the original nest with propolis on their hind legs. (The removal of pellets from the nest in the mandibles of workers is garbage removal, not removal of material to found a new colony.) The final stage is the flight of a virgin queen reared in the parent nest to the new nest. The new queen is chosen by the workers and escorted to the new site. She makes a mating flight from there, mates with a male from another colony, and returns to the new nest site to become the reigning queen.

Keepers of stingless bees have attempted to capture colonies by providing empty hive boxes. These attempts have had limited success. It seems that, despite attempts to lure scout bees that are searching for new nest sites, they do not generally recognise or accept any of the artificial hive boxes currently in use. (Read more about this in "Trap hives" in Chapter 10.)

The differing strategies of the two groups of highly social bees (Table 4-1) come with advantages and disadvantages. The stingless bee strategy appears safer, because they prepare the new nest site before they move to it. But this strategy limits stingless bee reproduction, because they can only found a new colony within the flight range of their original colony. Honey bees are able to exploit a much larger area — perhaps tens of kilometres. On the other hand, honey bee swarms sometimes perish without finding a suitable new nest site.

TABLE 4-1 # Colony founding of honey bees compared to stingless bees.

	Honey bees	Stingless bees
Departure of bees from the parent colony	Sudden, obvious	Gradual, inconspicuous
Preparation of the new nest	Not prepared in advance	Prepared in advance
Reigning queen	Flies off with the swarm	Cannot fly, stays in the old nest
New queen	Reared and remains in the original nest	Reared in the original nest but leaves to found the new colony
Contact between parent and new colonies	No contact after separation	Continues for weeks or months
Distance between parent and new colonies	Potentially large	Small (within flight range of about 500 m)

Note that founding a colony is different from **absconding**. Absconding refers to the relocation of a colony of bees from one nest site to a new one when conditions at the old location are unsuitable (e.g. due to lack of food, or build-up of pests or pathogens). Honey bees practise this, but stingless bees *cannot* abscond, they are fixed in their existing location, because their queen loses the ability to fly after her original mating flight. (Note that this inability to abscond is the key to understanding why stingless bees are not pacified by smoke. Read more about this in "Protecting yourself against the defences of stingless bees" in Chapter 13.)

To summarise this chapter, Table 4-2 compares and contrasts the nesting biology and reproductive biology of the highly social bees (honey bees and stingless bees).

TABLE 4-2 # Nesting behaviour of honey bees compared to stingless bees.

	Honey bees	Stingless bees
Provisioning of food for larvae	Progressive	Mass
Larvae develop in cells that are:	Open	Closed
Re-use brood cells	Yes	No
Opening of brood cells	On the side	Usually on top
Cells for food and brood	Equal	Different
Use of propolis	Light	Heavy
Storage of pollen	Small-scale	Extensive
Entrances	Open	Protected
Evaporative cooling	Yes	No
Metabolic heating	Yes	Partial
Absconding	Yes	No
Natural colony founding	Sudden, visible	Gradual, inconspicuous

5 Foraging behaviour of stingless bees

FAST FACTS

Stingless bees have a remarkable **ability to efficiently harvest food** from their environment.

Foraging stingless bees can **communicate distance and direction information** to other foragers. Unlike honey bees, they do this not through dances but by various other means.

Each stingless bee worker tends to **specialise in one** flower resource, but the colony as a whole opportunistically **exploits many** plant species.

Australian stingless bees **fly a maximum of 500 m** to their food sources.

Although a colony may be populated by thousands of workers, typically less than 1,000 are **foragers**.

In this chapter, we examine the remarkable abilities that stingless bees have evolved for foraging efficiently. You do not need to read this section to keep stingless bees, but it will improve your appreciation and understanding of your bees if you do. If you wish to use stingless bees for pollination, you will gain from knowing more about the behaviours associated with flower visits.

Finding food is not easy for any bee. Flowers come and go: different plants open their flowers at different times of the day and flower at different times of the year. Other competing animals may deplete the food. Some plants hide their food in complex flower structures. But bees are well equipped for the job. Bees use their senses, particularly sight and smell, to navigate to and from the nest and flowers. They use the position of the sun as a compass to help them get around and they possess a built-in clock. They memorise landmarks and retain a "map" of the area around the nest. They also have the ability to learn. Armed with these skills, bees efficiently work the area around their nest to harvest materials and transport them to the nest.

Activity at a stingless bee nest entrance: a returning forager laden with nectar, and a departing bee disposing waste.

IMAGE
JAMES DOREY

NECTAR

FIGURE 5-1 A stingless bee *(Tetragonula* sp.) sucking nectar from a lychee flower.
IMAGE **GIORGIO VENTURIERI**

5.1 How stingless bees collect their resources

Most stingless bees collect from their environment three main resources critical to their survival: **nectar**, **pollen** and plant **resin**. In Africa and the Americas, a few species live instead by robbing the nests of other species (social parasites), but in this chapter we consider only the non-parasitic species.

Nectar and pollen are collected from flowers. These are vital and complementary food sources: nectar provides sugary **carbohydrates** for energy, and pollen provides **protein** to build and repair bodies. Both foods may be consumed immediately for current needs or stored in the nest for future use. To be stored, the food is processed for preservation and placed in pots. Storage is important, because food may be abundant at some times of year but scarce at others. Stingless bees, like honey bees, take maximum advantage of an abundant source of food to collect it in volume. It is this tendency to store food for the hard times that provides the exquisite honey that humans crave.

5.1.1 NECTAR COLLECTION

Stingless bees find nectar mainly in flowers (Figure 5-1). The quantity of nectar in a flower varies enormously. For example, flowers adapted to pollination by bats produce huge quantities of nectar. Flowers visited by small bees typically produce less than 1 microlitre. This suits the small stingless bees. *Tetragonula carbonaria,* for example, can only carry about 2 microlitres back to the nest (see nectar-laden bee on page 64).

The sugar content of nectar also varies a lot, typically from 10% to 70%. When nectar is diluted, bees spend a lot of energy carrying it back to the nest and more energy dehydrating the nectar to make honey from it. In this way, bees may benefit from dry weather because the nectar is not diluted by rain.

5.1.2 POLLEN COLLECTION

Pollen is also collected from flowers, where it is produced in the **anthers** (the structure at the end of the stamen, a flower's male reproductive organ). (Read more about the structure of flowers in Chapter 15 "Bees and pollination".)

FIGURE 5-2 Stingless bees returning to nest with pollen loads. IMAGE **JAMES DOREY**

Bees have an impressive ability to learn a variety of handling skills to remove this granular substance from flowers. After collection, stingless bees may moisten the pollen with regurgitated nectar to make it stickier and easier to pack into their pollen-carrying area (Figure 5-2). Unlike nectar, pollen is not transferred from **foraging bees** to receiving bees near the nest entrance. Instead, the foraging bee carries the pollen directly to an open pollen storage pot. The forager may be offered nectar by **house bees** on returning from and leaving for a flight. This provides fuel for the flight, because pollen is low in carbohydrate.

Although the workers of *Tetragonula carbonaria* weigh only about 4 mg, their pollen loads weigh around 1.4 mg. The forager can carry this load for several hundred metres. This is equivalent to a human carrying 27 kg and flying about 100 km! By comparison, honey bees weigh 100 mg, about 25 times more; no wonder they can fly a lot farther, and carry far more pollen and nectar.

5.1.3 RESIN COLLECTION

Resins are substances secreted by plants to heal wounds. They are extruded as liquids but set hard with time, making them the ultimate nest-building material. At the plant, bees manipulate resin with their mandibles and pack it into pollen baskets on their hind legs to carry it back to the nest (Figure 5-3). You can see this as a usually clear liquid droplet on the hind leg of the bee. One well-known source of resin for stingless bees is found

FIGURE 5-3 Stingless bees collecting resin exuding from pine tree *(TOP)* and damaged eucalypt bark *(BOTTOM)*. IMAGE TOP **JAN ANDERSON**

inside the gumnuts (fruits) of cadaghi, a eucalypt named *Corymbia torelliana* (Figure 13-20). Back at the nest, the resin is stored for future use or used immediately. Rarely are Australian stingless bee species observed damaging plants to induce resin flows, although this is a common behaviour in some *Trigona* species in South America, which are infamous for damaging citrus trees.

5.1.4 WATER COLLECTION

Stingless bees do not use water for cooling their nests, but water may be collected by local stingless bees in some dry places and seasons. It is usually drawn from wet substrates, such as wet rocks near creeks, rather than from open water bodies where there is a danger of drowning. This water may be required for nutrition or to dilute honey for consumption. Alternatively, water may be collected for its salt content. In tropical areas, stingless bees are well known for collecting sweat from humans and so have earned the name "sweat bees".

5.2 Recruitment communication of foragers to resources

The famous communication dances of honey bees are one of the pinnacles of complexity of insect behaviour. In the round and waggle dances, scout bees communicate information to waiting foraging bees about the location of resources outside the nest. The resources may be flowers, resin sources, nest sites, etc. The dances take place on the comb in the nest in darkness, but from them the recipient bees learn the position of the resource, both its direction and distance from the nest. They can then fly directly to it. The new foragers are recruited to the resource by the information communicated to them by the scout bee.

Less is known about recruitment communication in stingless bees. Of the approximately 600 species of stingless bees, only about a dozen have been studied. However, reasonable progress is being made and understanding is increasing. Experiments are conducted by placing feeders with an attractive food source within flight range of a nest, marking foragers, and then observing what happens at the nest and the feeder.

Various species have different abilities to communicate the distance and direction of a food source. Some species cannot communicate distance or direction at all, but simply alert nestmates to the presence of a rewarding resource. Figure 5-4 shows this for the South American species *Plebeia droryana*. On the other hand, nine of the twelve studied species show a competent ability to communicate the correct distance and direction to nestmates (Figure 5-5), including height (three-dimensional spatial direction).

How do they do this? A complex dance has not been observed, but other mechanisms have been detected, including vigorous contact between the scout and the forager (jostling), sound (buzzing), vision (piloting of foragers by the scout to the resource) and smell (scent deposition and odour trails).

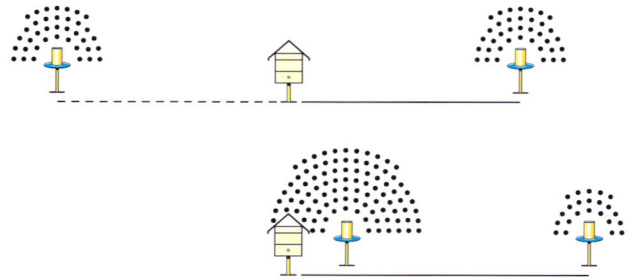

FIGURE 5-4 The results of two experiments on a species of stingless bee that has the simplest recruitment communication system (i.e. the scout bee alerts new recruits to a feeder but gives no distance and direction information). In both drawings, a hive is presented with two feeders and the feeder on the right has been discovered first. The top figure shows that, soon after the feeders are discovered, equal numbers of foragers (represented by the dots) arrive at both feeders, illustrating that no direction information has been communicated. In the next experiment, the bottom figure shows that, soon after the first feeder is discovered, more foragers arrive at the closest feeder in the same direction, confirming that no distance information has been communicated.

ARTWORK **GLENBO CRAIG** from M. Lindauer 1961.

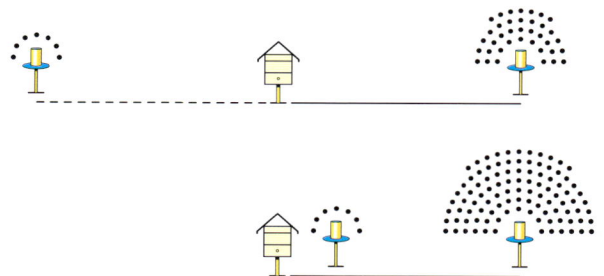

FIGURE 5-5 The results of two experiments on a species of stingless bee that has a complex recruitment communication system (i.e. the scout bee alerts new recruits to a feeder and gives accurate distance and direction information). In both drawings, a hive is presented with two feeders and the feeder on the right has been discovered first. The top figure shows that many foragers arrive at the feeder discovered first, illustrating that direction information has been communicated. In the next experiment, the bottom figure shows that more foragers arrive at the distant feeder in the same direction, confirming that distance information has also been communicated.

ARTWORK **GLENBO CRAIG** from M. Lindauer 1961.

The recruitment communication system of just one Australian species has been investigated. Tad Bartareau studied *Tetragonula carbonaria* and observed that they leave a scent mark on the artificial feeders. Tad then teamed up with world expert on bee communication, James Nieh, to show that *T. carbonaria* is able to recruit nestmates to food. But the distance and direction information is not as complete as that communicated by honey bees and many South American stingless bees. *Tetragonula carbonaria* bees are able to communicate direction but not distance (Figure 5-6). This bee relies heavily on flower scent to find food and then the same scent to recruit foragers to the food. In addition, it uses a marking pheromone deposited by experienced foragers that then attracts new recruits (Figure 5-7).

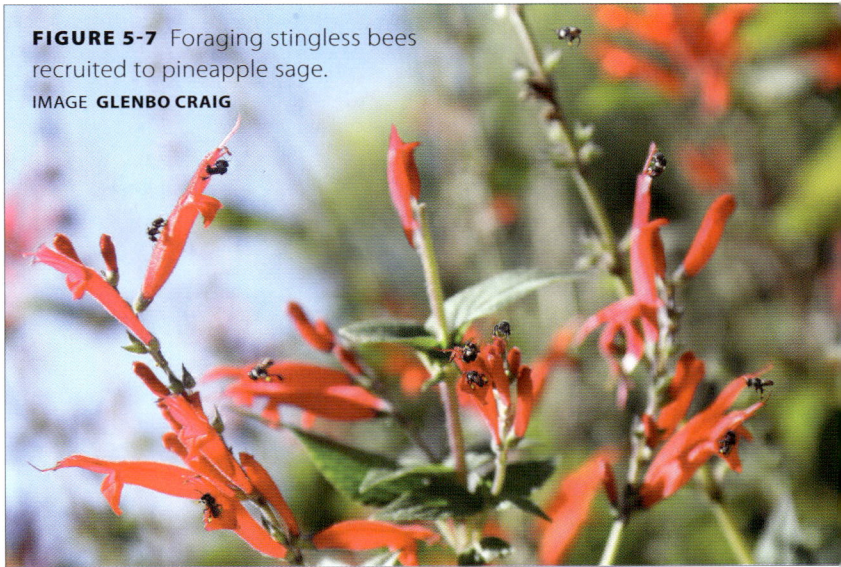

FIGURE 5-7 Foraging stingless bees recruited to pineapple sage.
IMAGE **GLENBO CRAIG**

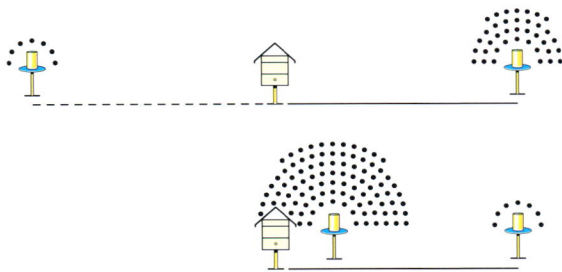

FIGURE 5-6 The results of two experiments on the stingless bee species *Tetragonula carbonaria*, which has a hybrid recruitment communication system (i.e. a scout bee alerts new recruits to a feeder and gives accurate direction but not distance information). In both drawings, a hive is presented with two feeders and the feeder on the right has been discovered first. The top figure shows that many foragers (represented by the dots) arrive at the first-discovered feeder, showing that direction information has been communicated. In the next experiment, the bottom figure shows that more foragers arrive at the close feeder in the same direction as the first discovered feeder. This confirms that no distance information has been communicated and that the new recruits headed in the right direction and found the first source in that direction.
ARTWORK **GLENBO CRAIG**

The ability of *Austroplebeia* species of bees to recruit nestmates to flowers is not well known, but it appears from Megan Halcroft's studies on their foraging behaviour in a greenhouse that they have a lesser ability than *T. carbonaria*. *Austroplebeia* foragers demonstrated solitary foraging strategies, while *T. carbonaria* showed constant, high recruitment of foragers and group foraging strategies. In contrast, Anne Dollin observed *Austroplebeia* foragers in central Australia marking artificial feeders with a fluid, presumably a marking pheromone to help others find the resource (Figure 5-8).

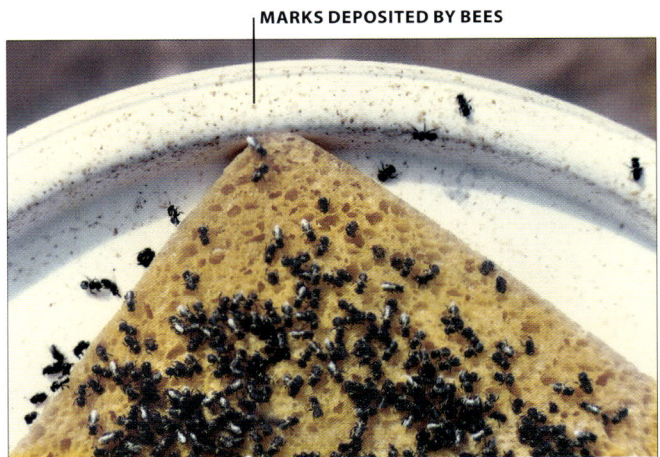

MARKS DEPOSITED BY BEES

FIGURE 5-8 These foragers of *Austroplebeia* in central Australia have marked the artificial feeder with a marking pheromone.
IMAGE **ANNE DOLLIN**

5.3 Foraging strategies

Different species of stingless bees use a variety of strategies to collect resources from the environment. Most species of stingless bees are generalists and opportunists. This means that colonies are able to exploit the pollen and nectar of many plant species. They also readily adapt to introduced plant species, including crop species.

This aptitude for using the pollen of many plant species is called **polylecty**. But polylecty at the colony level is balanced by **floral constancy** at the level of the individual; a forager on a trip usually only visits one plant species for one particular resource. Sometimes, while collecting pollen, the bee will take a sip of flower nectar to provide the energy she needs to get back to the nest. But this is just incidental to the task of collecting pollen. This is not to say that a bee will dedicate itself for its entire life to a particular resource. On the contrary, she will normally switch to other foraging tasks as resources vary from one day to the next, and even within a day.

Keepers of Australian native bees have long pondered the different foraging behaviour of the two genera of Australian stingless bees, *Tetragonula* and *Austroplebeia*. Recently, I worked with Sara Leonhardt and Helen Wallace to study these differences in detail, and we confirmed that the differences are indeed pronounced. We analysed the resource intake and foraging activity of *Tetragonula carbonaria* and *Austroplebeia australis* at the same site within their natural distribution in southern Queensland. We found that *T. carbonaria* had higher activity (about 60 foragers per minute) than *A. australis* colonies (about 15 foragers per minute). *T. carbonaria* also showed a higher sugar intake per minute and collected more resin and more pollen from a broader range of plant species. In contrast, the smaller *A. australis* colonies collected from fewer types of plants. Instead, they focused on high-quality resources (i.e. nectar of higher sugar concentrations). Their foraging activity was highly variable from day to day.

5.4 Flight range

The flight range of stingless bees correlates closely with their size. This relationship was determined in studies on Central American stingless bees, and predicts a flight range of a little less than 500 m for our smaller Australian species, and a little over 500 m for our larger ones. This was confirmed in a recent homing study by Jordan Smith, who marked foragers of the intermediate-sized *Tetragonula carbonaria* bees, released them from increasing distances, and showed that most returned home at distances less than 500 m, but few returned home at distances greater than 500 m. This **homing range** is probably a good estimate of the actual foraging distance of bees, but it is their maximum range and they will ordinarily forage closer to their nest than this.

It is very useful to know bees' flight range. It allows beekeepers to assess the suitability of a position for hives. Try drawing a 500 m radius around where you wish to place your hives and observe the available landscape (Figure 5-9). The presence of flowering trees and gardens is desirable, while grassland, conifer forests, hard surfaces and water are not. Knowledge of the flight range also allows us to know how far to move hives without fear of foragers returning to the old location. In addition, the flight range sets the boundaries for where an established colony of stingless bees can found a new colony.

FIGURE 5-9 500 m radii around two hive sites.
Around a school in an urban area, a mix of gardens and parks is available as forage for colonies of bees (LEFT). Around a farmhouse, a mix of forest and tree crops is available (RIGHT).
IMAGE http://www.freemaptools.com/radius-around-point.htm

MEASURING FLIGHT RANGE

Researchers can measure flight range in at least four ways.

1 Capture workers, mark them, release them at increasing distances from the nest and see if they return. This is more correctly the homing range. This method was used by Jordan Smith to estimate a range of 500 m for *T. carbonaria*.

2 Train workers to collect from an artificial nectar source and progressively move the source away from the nest. Using this method, the maximum flight range of some tropical American species was found to range from 120 m to 980 m. This method is subject to conditions, because foragers will travel farther from a colony if they have inadequate food reserves and if they are in a poor environment with few food alternatives.

3 Use the calculated flight speed and time spent away from the nest. This measures the real distances normally travelled, rather than the maximum range, but is imprecise. Using this method, an Asian species was shown to forage in the range of 100 m to 400 m.

4 Measure the presence of bees on farms away from natural vegetation where they nest (e.g. an Asian species foraged on oil palms in a plantation 1 km from the forest they inhabited). Also, the abundance of one stingless bee species in a longan orchard in northern Thailand was high at distances of 50 m and 200 m from adjoining forest, but decreased greatly at 2.5 km and 4 km from the forest.

5.5 Foraging population

The total number of bees foraging from a hive is a useful statistic. Giorgio Venturieri and I recently assessed this for hives of *Tetragonula carbonaria* situated on farms in south-east Queensland.

Towards the middle of sunny days in August and September, we closed the hive entrance and collected the returning foragers. We counted the number returning in 10-minute intervals until no more bees returned, usually by about 60 minutes. The collected bees from each period were held in separate containers and not released until the end. We found an average of approximately 1000 foraging bees per hive, with a range of 500 to 1700. This figure is an underestimate of the foraging force of a colony, as some foragers will be trapped in the hive upon closing the entrance. Figure 5-10 shows a typical case in which 1,215 foragers returned to the hive over 80 minutes. The bees returned quickly at first but, as all the foraging bees returned and none could leave the closed hive, the arrivals quickly dropped off. The median time of return was 18 minutes.

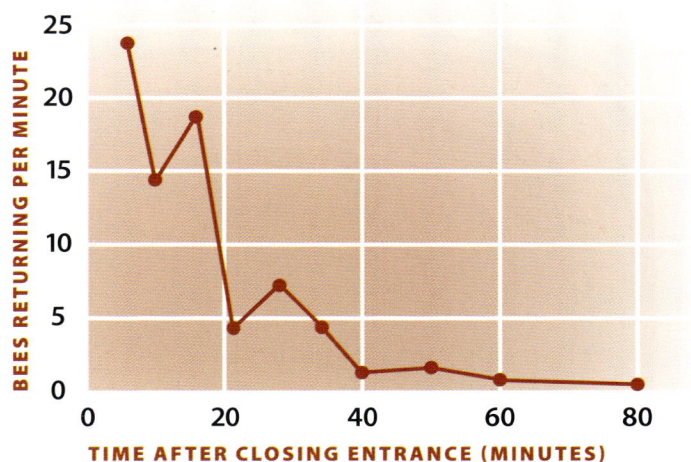

FIGURE 5-10 Rate of return to a hive of *Tetragonula carbonaria* foraging bees on a farm in south-east Queensland on a sunny winter day.

Foraging bees make many flights per day, so the total number of flights from a colony of bees in a day is much higher than 1,000. We estimated that each hive achieved about 20,000 flights per day. This number will vary, depending on factors including the population of the colony, the needs of the colony and the local availability of resources. Colonies that are larger, with few stored provisions and with access to good flowering, tend to make more flights.

Giorgio Venturieri and members of the indigenous Palikur, inspecting a nest of *Melipona eburnea* stingless bees in the north of Brazil. IMAGE **MARCIO SZTUTMAN**

6 Global diversity and distribution of the highly social bees

Honey bees consist of about 10 species in the Old World, whereas there are around **600 species of stingless bees** found throughout the world's tropics.

The Amazon rainforest is the centre of stingless bee diversity; in comparison, Australia is poor in species.

Stingless bees **evolved in Africa and South America** when these two continents were connected, and took millions of years to reach Australia through Europe and Asia.

Tasting honey from stingless bees in a log hive, East Africa.

Honey bee distribution

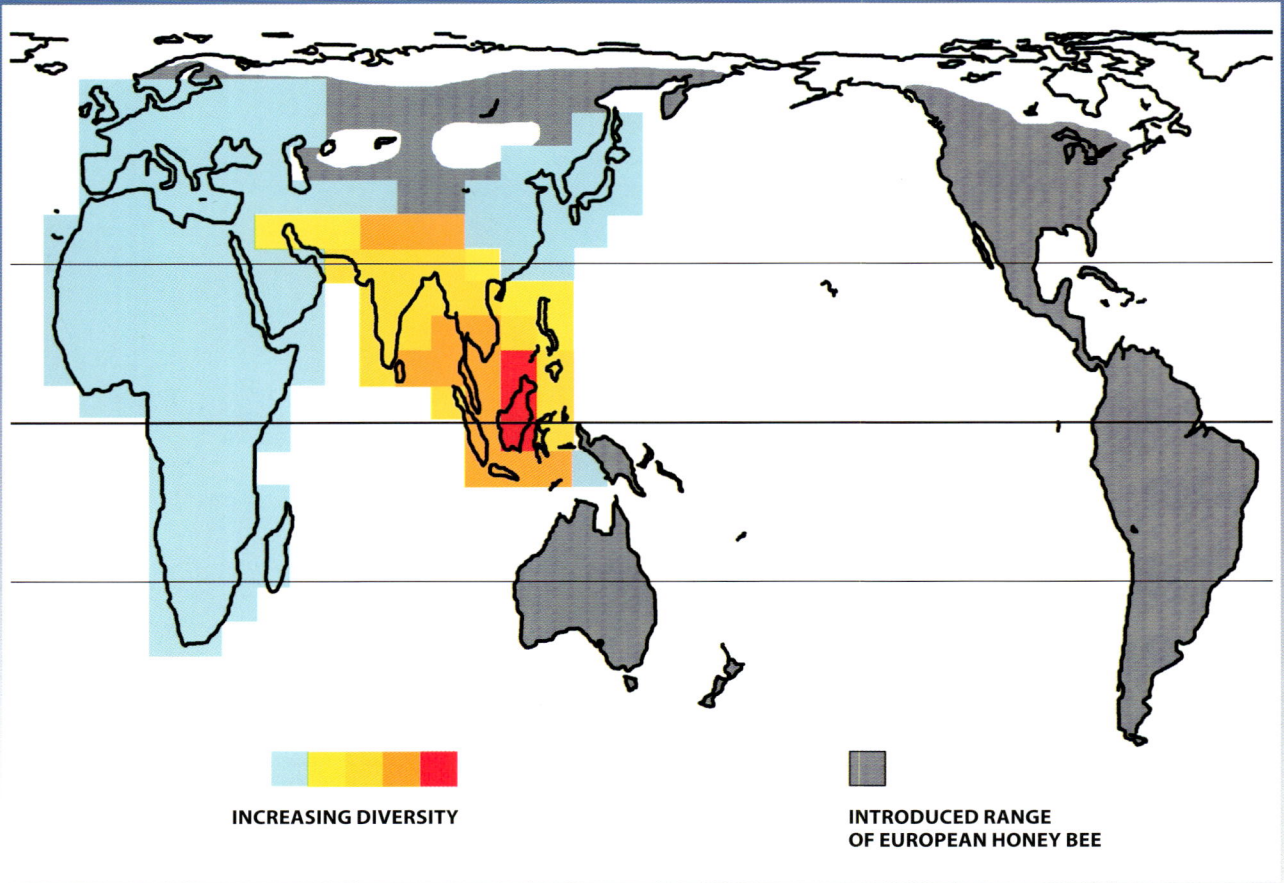

INCREASING DIVERSITY

**INTRODUCED RANGE
OF EUROPEAN HONEY BEE**

FIGURE 6-1 Diversity and global distribution of the honey bees. The red colour shows high numbers of species and the pale blue indicates few. The grey tonal areas show where the European honey bee has been introduced for apiculture. IMAGE WITH PERMISSION OF UK NATURAL HISTORY MUSEUM. REDRAWN BY **TOBIAS SMITH**

6.1 Diversity and distribution

The two highly social tribes of bees, the honey bees (Apini) and the stingless bees (Meliponini), are sister groups (Figure 2-1), but they separated many millions of years ago. These two groups, sometimes referred to as the apines and meliponines, have had plenty of time to spread widely and evolve into new species. They now occupy large expanses of the globe. The two groups show quite different patterns of diversity and distribution.

The honey bees have only one genus, *Apis,* which contains only about 10 species. Its distribution is in the Old World, with species in Eurasia and Africa (Figure 6-1). The name "European honey bee", *Apis mellifera,* is actually a misnomer, as

this species occurs from the north of Europe to the southern tip of Africa. This species consists of several **races**, for example, the tropical African race, *Apis mellifera scutellata*, and the Italian bee, *Apis mellifera ligustica*. Italian bees, or hybrids of them, are the most common subspecies kept for honey production and pollination in Australia. The other species of *Apis*, such as *Apis cerana*, the Asian honey bee, are native to Asia.

South-East Asia is a hotspot of diversity of *Apis*. It is curious that no species of *Apis* made the crossing and established in Australia, but that the stingless bees did so. (Read more about this in "Biogeography of Australian stingless bees"

Stingless bee distribution

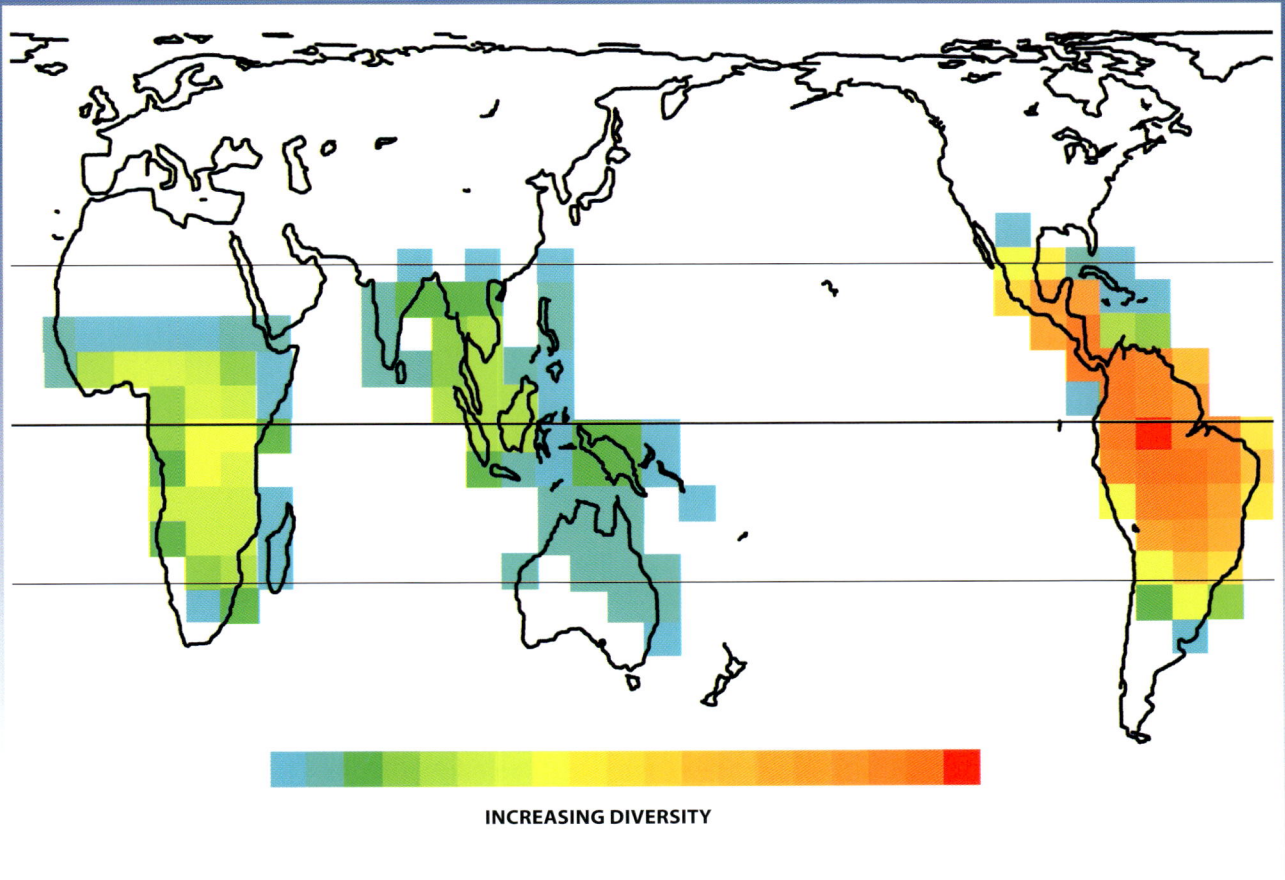

FIGURE 6-2 Global distribution and diversity of the stingless bees. The red colour shows high numbers of species and the pale blue indicates few.
IMAGE WITH PERMISSION OF UK NATURAL HISTORY MUSEUM
REDRAWN BY **TOBIAS SMITH**

below.) It is intriguing to imagine that, instead of stingless bees, *Apis* honey bees could have migrated to Australia and become native here. Perhaps it is just by chance that Australia is naturally occupied by stingless bees and not Asian honey bees. Or perhaps stingless bees have a superior ability to cross the seas in their closed nest within a floating log. The open nests of honey bees are less likely to float.

The stingless bees are much more diverse than the honey bees, consisting of about 50 genera and an estimated 600 species. Their distribution is pan-tropical, as they are found in most tropical mainland parts of the world. Their diversity is uneven, peaking in the Amazonian rainforest but relatively low in Australia (Figure 6-2).

About 400 species of stingless bees are known in the Americas, approximately 25 species in Africa and Madagascar, about 80 in South-East Asia, and 11 in Australia. The diversity of species in the Americas is matched by the variety of their shape, size, colour, ecology and behaviour. The example of size range underlines this point. In Australia, our stingless bees are relatively uniform in size, with the workers of various species differing in length from approximately 3 mm to 4.5 mm. But the Americas boast species that are larger than the European honey bee.

FIGURE 6-3
Tetragonisca angustula,
a common stingless bee
from the Americas.
IMAGE **TOBIAS SMITH**

Although this book is primarily about Australian stingless bees, one cannot ignore the stunning richness of form and behaviour of the American species. To give a taste, I mention just two species. *Tetragonisca angustula* (Figure 6-3) is commonly kept in hives over a huge range from Mexico to Argentina. This small but striking bee is common in towns, where they nest in cavities in human structures. The honey of this species is excellent and is highly prized for food and medicine.

The most popular bees kept in the Americas belong to the *Melipona* genus. *Melipona* is a distinct genus of stingless bees that are generally large in size, have a unique form of queen production and an ability to buzz pollinate. In Amazonian Brazil, *Melipona fasciculata* is a common species that is kept in hives for honey production (Figure 6-4).

Stingless bees are conspicuously absent from isolated islands. These social insects cannot live as individuals but only as a colony housed in a protected place, usually a hollow log. So, for them to migrate across seas, the entire nest has to float. This must be a rare event and so they never reached many small and isolated islands, such as most Pacific islands.

FIGURE 6-4 A colony of *Melipona fasciculata* in a hive box in Brazil.
IMAGE **GIORGIO VENTURIERI**

6.2 Biogeography of Australian stingless bees

Stingless bees are found in all tropical parts of the world. How did they come to be so widely distributed? How and when did the two groups (the genera *Tetragonula* and *Austroplebeia*) that occur in Australia arrive here? The answers lie in the long history of these bees (they are about 80 million years old), the movement of the land masses of the Earth (**continental drift**), the separation and isolation of the **biota** (the total collection of bee species of a region or a time), the point at which continents broke up (called **vicariance**), the changing global climate over tens of millions of years, and the migration of species when the climate was suitable (**dispersal**). Read on and be astonished by the long journey our stingless bees took before finally settling here and becoming part of Australia's native fauna. Brace yourself for the final twist when it is revealed that the two groups arrived here by different paths!

A convincing theory that explains the present distribution of the stingless bees was recently proposed by Claus Rasmussen and Sydney Cameron. They used the modern technique of molecular systematics to arrive at a family tree of all the world's stingless bees. They collected fresh specimens of hundreds of species, then extracted and sequenced their DNA. By comparing the differences in the DNA sequences, they produced a family tree (a phylogeny) that provides a foundation for understanding how stingless bees come to be where they are today.

The first insight that arose from their studies was that the family tree of stingless bees is divided into two major branches: the New World branch (species from the Americas) and the Old World branch (from Africa, Asia and Australia). The Old World species are more closely related to each other than they are to species from the New World and vice versa. This suggests that the original stingless bees split into two isolated groups that occupied these two areas.

Another fascinating aspect is that Rasmussen and Cameron were able to date when some of these events occurred. They used fossils of stingless bees to estimate when splits occurred in the evolution of these bees. The estimate for the split between the Old World and New World branches of the tree is about 70 million years ago. It is not a coincidence that the ancient supercontinent of Gondwanaland was breaking up around this time. Antarctica and Australia had already drifted off. What is now South America was breaking away from the land mass that formed Africa (Figure 6-5). The original stingless bees that occupied this great southern land were torn apart, never to meet again.

70 million years ago

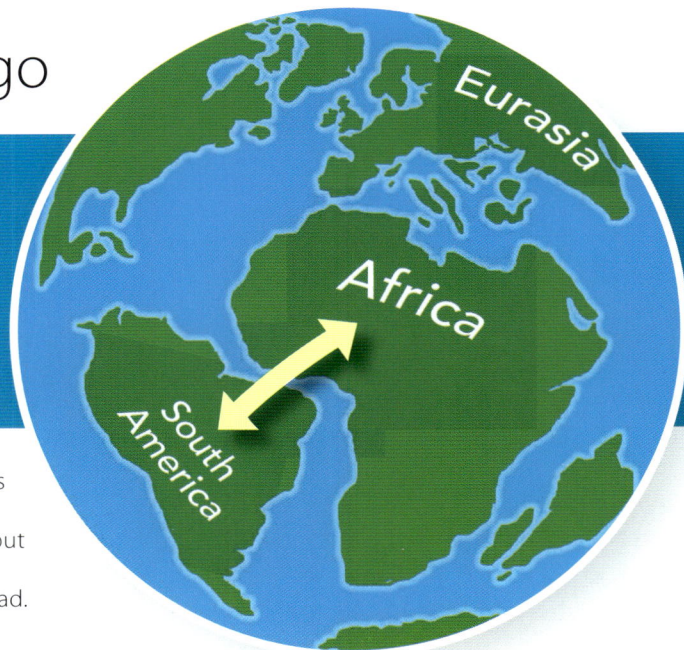

South America

Africa

FIGURE 6-5 The oldest split in the stingless bee family tree caused by geological separation of Africa and South America about 70 million years ago. The separation of the bees is represented by the double arrowhead.
ARTWORK **TOBIAS SMITH** AND **GLENBO CRAIG**

The bees in South America flourished and evolved into many new species as their homeland drifted to its current position. They were completely isolated and diverged from the old lineage in Africa to the point that they can now be recognised as different branches of the tree. When South America drifted farther north and collided with North America, the bees colonised the newly accessible territories of Central America.

Meanwhile, Africa continued its journey north and collided with Eurasia about 50 million years ago, forming a land bridge for the stingless bees to spread across into present-day India and South-East Asia (Figure 6-6). The climate was warm and humid at this time and so the stingless bees also spread into Europe.

50 million years ago

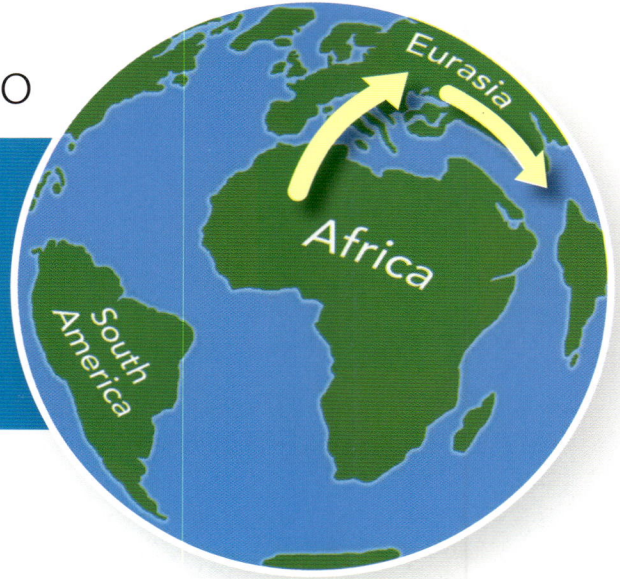

FIGURE 6-6 Stingless bees dispersed into Eurasia when Africa collided with that land mass about 50 million years ago. The dispersal of the stingless bee biota is represented by the single arrowhead.
ARTWORK TOBIAS SMITH AND **GLENBO CRAIG**

Later, conditions in the Middle East became unfavourable for survival of these bees, possibly as the climate cooled and dried. The bees in Asia were on their own, separated from their forebears in Africa. This explains the next major branching of the evolutionary tree. This branch separates the African from the Asian species (Figure 6-7).

38 million years ago

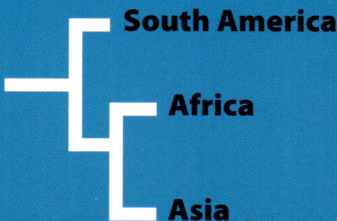

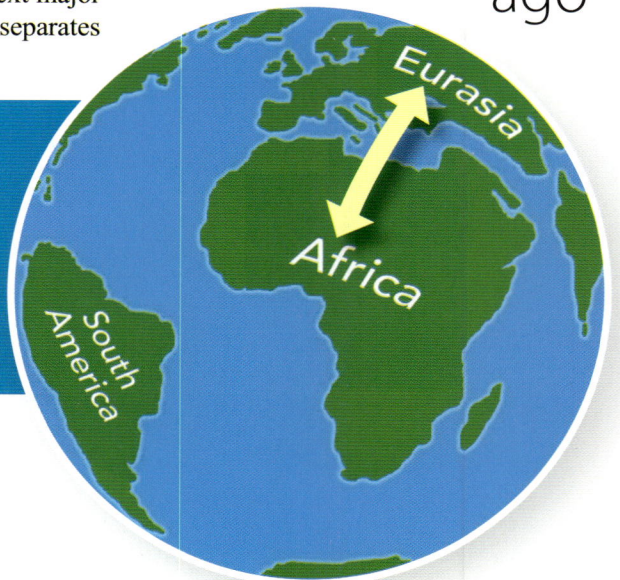

FIGURE 6-7 The next major split in the stingless bee family tree caused by the separation of the Asian and African clades. The separation is represented by the double arrowhead.
ARTWORK TOBIAS SMITH AND **GLENBO CRAIG**

Finally, the stingless bees arrived in Australia from Asia (Figure 6-8). We know this because the Australian species are closely related to the Asian species. We also know that there were probably several invasion events. The *carbonaria* group, which includes the common species *T. carbonaria*, *T. hockingsi* and *T. mellipes*, is the most distantly related to the other Australian and Asian *Tetragonula*. It is probable that this was the first invasion of a common ancestor of this group from Asia to Australia, and these bees then evolved to form the species that exist today. The Asian ancestor may be *Tetragonula biroi*, which is the only species of the *carbonaria* group in Asia. The north Queensland species *T. clypearis* and *T. sapiens* also occur in Asia and clearly arrived separately and later as they have not had enough time to become distinct species.

In all, it appears that there were three dispersal events of *Tetragonula* to Australia. The frequency of this event is due to the periodic land bridges that have spanned Cape York and Papua New Guinea, the last one being broken as recently as 10,000 years ago. Land bridges form when sea levels fall because water is locked up in ice at the poles during ice ages.

Sometime between 1,000,000 and 10,000 years ago…

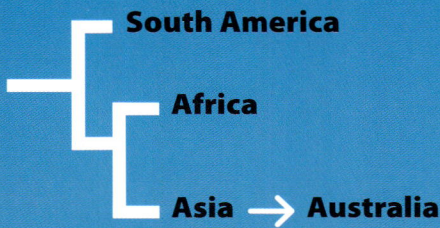

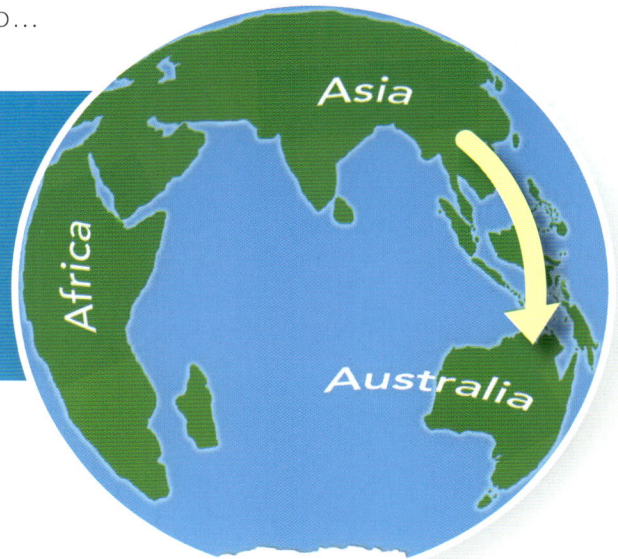

FIGURE 6-8 Stingless bees dispersed into Australia as that continent approached South-East Asia. The dispersal is represented by the single arrowhead.

ARTWORK **TOBIAS SMITH** AND **GLENBO CRAIG**

FIGURE 6-9 Fossil of the extinct stingless bee *Liotrigonopsis rozeni* in amber is about 45 million years old and proved valuable in dating the timing of events in the evolution of stingless bees.

Image courtesy Michael S. Engel, University of Kansas.
From Engel, M.S. 2001. "A monograph of the Baltic amber bees and evolution of the Apoidea (Hymenoptera)". *Bulletin of the American Museum of Natural History* 259: 1-192.

How do they know this stuff?

How can we know what happened all those millions of years ago? Modern genetic techniques provide key evidence, as does the fossil record. Luckily, stingless bees have left a long record in amber (fossilised tree resin). For example, an amber fossil of *Cretotrigona prisca* has been dated to approximately 65 million years ago; the age of the stingless bees must therefore be greater than this.

But there's a twist.

The *Austroplebeia* genus does not appear to follow the timeline described above for *Tetragonula*. *Austroplebeia*, along with the Asian *Lisotrigona*, belongs to the African branch of the tree, not the Asian one. The closest relative of *Austroplebeia* and *Lisotrigona* is the African *Liotrigona* (yes, *Liotrigona* as opposed to *Lisotrigona* – confusing, isn't it?). It appears that there was a dispersal from Africa to Asia and then to Australia, well after the isolating event of the separation of Africa from South America, and also after the isolation of the African and Asian elements (Figure 6-10). This dispersal event happened independently to that of the bees that developed into *Tetragonula*. The ancestors of *Austroplebeia* left Africa later than those of *Tetragonula*. But they arrived in Australia long enough ago to become so different that they now qualify as a genus in their own right.

Sometime between 1,000,000 and 10,000 years ago...

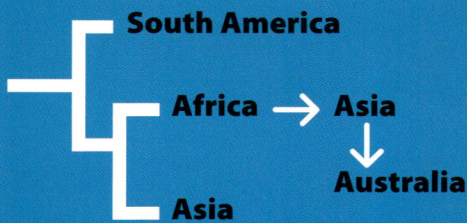

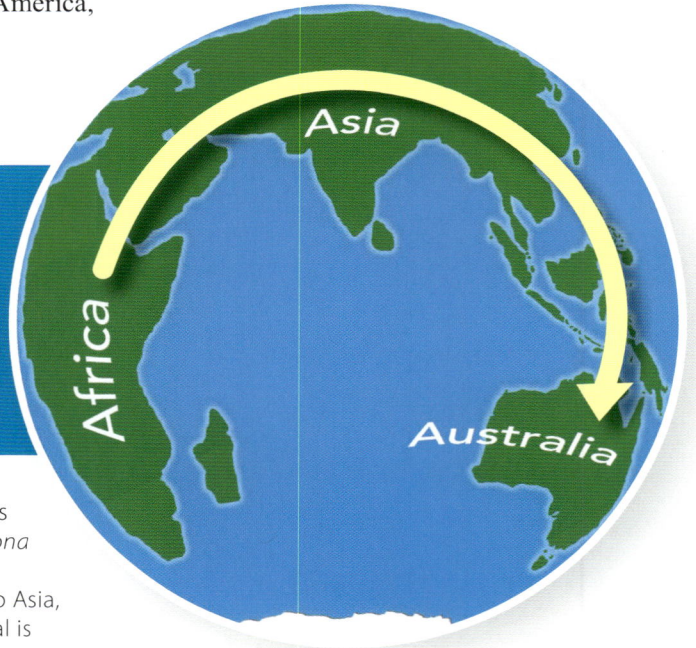

FIGURE 6-10 One group of stingless bees (*Austroplebeia* from Australia and *Lisotrigona* from Asia) belongs to the Africa clade of stingless bees; they probably dispersed to Asia, then Australia more recently. The dispersal is represented by the single arrowhead.
ARTWORK **TOBIAS SMITH** AND **GLENBO CRAIG**

In summary

Stingless bees are a Gondwanan group that formed on the South American/African landmass after it separated from Australia. As the biota of South America and Africa were forever separated and experienced no further gene flow between them, they formed two distinct clades. The bees then dispersed along a land bridge created by the colliding of Africa with Eurasia. Climatic change caused a separation and isolation of the African and Asian biota. The Asian biota then dispersed to Australia when Asia and Australia collided. Therefore, our Australian *Tetragonula* are closely related to other species in their genus in South-East Asia. *Austroplebeia*, however, is an African clade that arrived here from Africa after the separation of the African and Eurasian biotas.

Now that we have explored the stunning global diversity of stingless bees and accompanied them on their inter-continental journey over millions of years to reach Australia, let's discover the species that currently exist on our shores.

7 The stingless bees of Australia

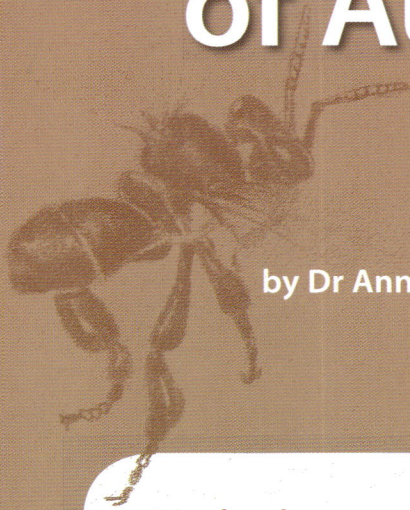

by Dr Anne Dollin and Dr Tim Heard

FAST FACTS

Stingless bees can be distinguished from other native Australian bees by their pollen baskets, small size, and fewer veins in their wings.

The eleven species of Australian stingless bees are placed in two genera, *Tetragonula* and *Austroplebeia*.

Australia has six species of **Tetragonula,** two of which we share with Asia. Many other *Tetragonula* species are found in Asia.

The genus **Austroplebeia** has five species and is endemic to Australia and New Guinea.

Nests and workers of the two genera can be distinguished with a little effort, but separating species within both genera is extremely challenging.

Although the **workers of different species** do not vary much in size, their nests vary from an average of 1 litre up to 10 litres in volume.

The distribution and nest architecture of the eleven species are now moderately well known.

A nest of the stingless bee *Austroplebeia essingtoni* in an opened log in the Kimberley, Western Australia.

7.1 How to distinguish stingless bees from other bees

In Chapter 1 ("Bee basics"), we learnt what makes bees different from other insects.

We start this chapter by exploring how stingless bees differ in appearance from other bees, with the goal of helping you to identify them.

The first obvious difference is that female workers of the Australian stingless bees possess a corbicula, or pollen basket, on each hind leg. This is a smooth, concave area used for carrying balls of pollen (Figure 1-7, Figure 3-5, Figure 7-1). (Note that only workers, not males, have a corbicula. For more tips on how to distinguish worker from male stingless bees, see "How to distinguish castes" in Chapter 3.)

The only other bees to have a corbicula are honey bees, bumble bees and orchid bees, none of which are native to Australia. Two species of honey bee and one species of bumble bee have established in Australia since European settlement, but they are easily distinguished from native stingless bees. At either 10 mm or 12 mm long, the two species of introduced honey bee are much larger than Australian stingless bees, which are all less than 4.5 mm long. Introduced bumble bees are also larger (8–25 mm long) and have so far invaded only Tasmania, which is well outside the geographic range of stingless bees.

To confirm your identification of a stingless bee, check the front wing. Generally, stingless bees have fewer veins and be can be differentiated from all other bees by the lack of certain cross-veins on the front wing (Figure 7-2).

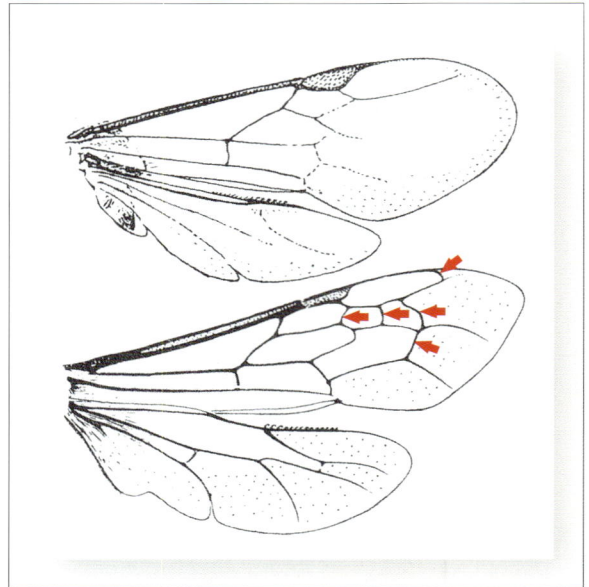

FIGURE 7-2 Wings of a stingless bee (*Austroplebeia*) *(ABOVE)* and a typical bee (*Leioproctus*) *(BELOW)*; the arrows mark the veins that are absent in the stingless bees.

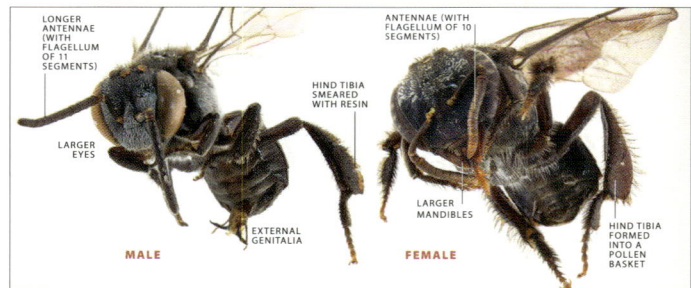

For a revision of external sex differences between male and female stingless bees, see this image, enlarged and explained, in Chapter 3, Figure 3-7.

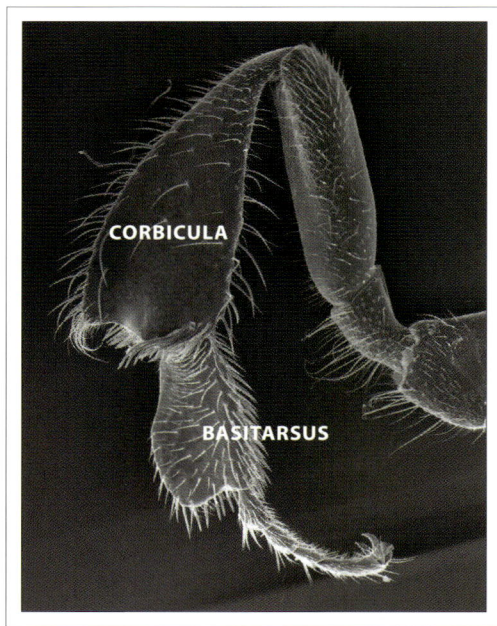

FIGURE 7-1 The pollen basket, or corbicula, on the hind leg of *Austroplebeia australis*.
IMAGE **MEGAN HALCROFT**

7.2 Comparing the two Australian genera:

Tetragonula and *Austroplebeia*

The stingless bees of Australia are not as diverse as those in parts of the world that have extensive tropical rainforests, but our bees do have some intriguing characteristics.

Only two of the world's fifty stingless bee genera or subgenera occur in Australia: *Tetragonula* and *Austroplebeia*. They belong to two separate branches within the tree of evolution of the stingless bees. (Read more about this in "Biogeography of Australian stingless bees" in Chapter 6.) They have been separated without interbreeding for long enough that they have evolved many differences. Table 7-1 contrasts the genera.

IMAGE **JENNY THYNNE**

IMAGE **BERNHARD JACOBI**

TABLE 7-1 A summary of the differences between the two genera of stingless bees in Australia.

	Tetragonula	*Austroplebeia*	More information
Number of species	6 known	5 known	Table 7-2
Biogeographic origin	Asian	African	Chapter 6
Distribution	In Australia and Asia	Endemic to Australia and New Guinea	Chapter 6 and Figure 7-3
Climate	Mainly in high-rainfall areas	In both high- and low-rainfall areas	Figure 7-3
Body colour	Black head and thorax	Black with pale markings on face and thorax	Figure 7-4
Back edge of thorax	Projecting	Rounded	Figure 7-4
Inside tibia of hind leg	Hairy ridge narrow	Hairy ridge broad	Figure 7-5
Nest construction materials	Heavy use of propolis	Light use of propolis, liberal use of wax	Figure 7-6
Nest entrance	Always open (but see *T. sapiens* and *T. clypearis*)	Closed at night and other periods of inactivity (but see *A. cincta*)	Figure 7-7 (see exceptions in text)
Brood architecture	Variable: regular comb, semi-comb or cluster	Cluster (but see *A. cincta*)	See images in species sections below
Brood cell orientation, shape and position of opening	Vertical, larval cells and cocoons oval, always open on top.	Irregular, larval cells spherical, cocoons oval, open in any direction: upwards, sideways, or downwards	For *Tetragonula* see: Figure 7-11 and Figure 7-15. For *Austroplebeia* see: Figure 7-8, Figure 7-9 and Figure 7-37
Laying of eggs by workers	No	Occasionally	Read more in Section 3.3.3 Worker reproduction

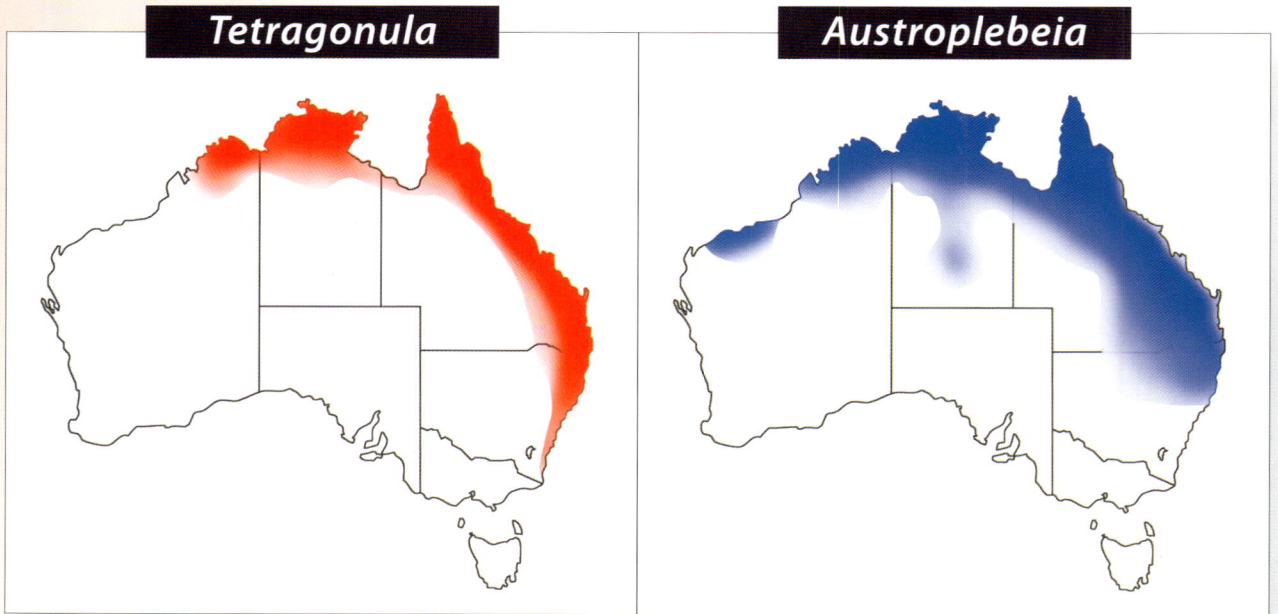

FIGURE 7-3 Estimated current distributions of *Tetragonula* (red) and *Austroplebeia* (blue) stingless bees.

FIGURE 7-4
Comparison of the body colouring and shape of the thorax of *Tetragonula* vs *Austroplebeia*.
A: A black scutellum with a projecting hind margin in *Tetragonula*.
B: A scutellum with cream markings and a rounded hind margin in *Austroplebeia*.
IMAGES **TOBIAS SMITH**

The inside of the tibial segment of the hind leg (the opposite side of the corbicula) also distinguishes *Tetragonula* from *Austroplebeia*. Both genera sport a ridge of distinctive dense hairs of uniform length (called keirotrichia) used for cleaning the wings. *Tetragonula* has a narrow ridge, whereas *Austroplebeia* has a broad area (Figure 7-5).

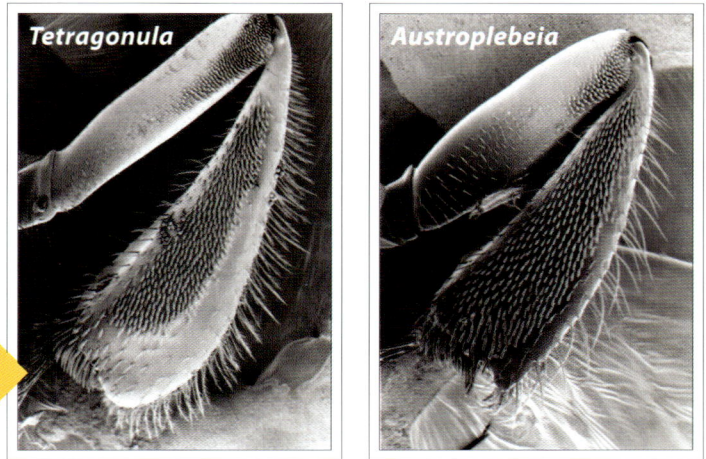

FIGURE 7-5 The inside of the tibial segment of the hind leg distinguishes *Tetragonula* from *Austroplebeia*. *Tetragonula* (LEFT) has a narrow ridge of distinctive dense hairs, whereas *Austroplebeia* (RIGHT) has a broad area.
ELECTRON MICROSCOPE IMAGES **BRONWEN CRIBB**

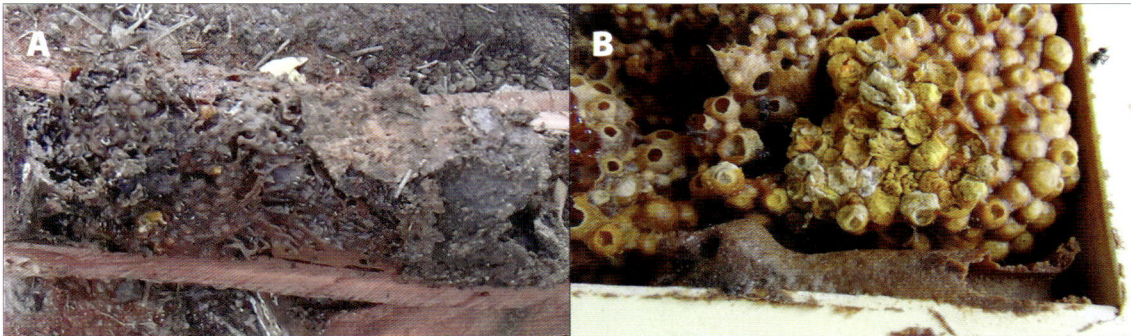

Tetragonula tend to build all nest structures from propolis, whereas *Austroplebeia* generally construct their nests largely from wax (Figure 7-6).

In most areas, the nest entrance provides a quick and easy way of distinguishing the two genera. *Austroplebeia australis*, *A. cassiae*, *A. essingtoni* and probably also *Austroplebeia magna* construct a fine, lacy curtain of resin droplets over their nest entrance at night or in other periods of inactivity. In contrast, *T. carbonaria* and *T. hockingsi* leave their nest entrances open at night. Anne and Les Dollin recently found that *T. clypearis* and *T. sapiens* bees may also partially close their entrance tunnels at night with a curtain that can be quite similar to that of the *Austroplebeia* (Figure 7-7). These sticky curtains protect against ants and other predators.

Austroplebeia cincta colonies in Queensland often do not close the entrance and, in the nests that do, the closure is not a fine lacy curtain but a semi-solid wall. In his classic 1961 publication on Australian and New Guinea stingless bees, bee diversity specialist Charles Michener said that *A. cincta* in New Guinea do close their entrances.

FIGURE 7-6 Nest structures of *Tetragonula* compared to *Austroplebeia*. **A:** *Tetragonula carbonaria* nest in a log showing the dark nest wall and food pots made of propolis rich in black plant resin. **B:** *Austroplebeia australis* nest showing the pale colour of entrance tube and pollen and honey pots, indicating that they are made of a propolis that is rich in wax with only a little resin.
B IMAGE **CLAUDIA RASCHE**

FIGURE 7-7 Closed nest entrances of *Tetragonula clypearis* (LEFT) and *Austroplebeia cassiae* (RIGHT) at night. IMAGES **ANNE DOLLIN**

FIGURE 7-8 Cocoons in the brood comb of *Austroplebeia essingtoni*. Note that the cells are not all upright but oriented in various directions.

FIGURE 7-9 The cluster brood comb of *Austroplebeia australis*. Note that these larval cells are spherical and that the two open cells are at different stages of completion because construction is continuous rather than batched. IMAGE **JEFF WILLMER**

Another interesting difference lies in the rate at which brood cells are constructed and queens lay eggs. Workers of *A. australis* and *A. cassiae* build cells continuously and the queen lays her eggs at a gradual pace, one at a time (Figure 7-9). In contrast, *T. carbonaria* and *T. hockingsi* workers build cells synchronously in batches of up to 90 brood cells at a time, about five times a day, and the queen lays her eggs in a rapid frenzy when the cells are provisioned (Figure 7-11, Figure 7-15).

The body length of Australian stingless bees does not vary much (Table 7-2). But *Austroplebeia* are stouter and heavier than *Tetragonula*. For example, although workers of *A. australis* and *T.*

carbonaria are about the same length, the former weighs 6.5 mg while the latter is considerably lighter at 4.5 mg.

Most species of Australian stingless bees can be successfully transferred to hives and artificially propagated. The colony benefits from a hive of appropriate volume. See the nest volumes in Table 7-2, and choose either a standard sized hive (for species with a nest volume of 5 litres or greater), or mini hive (for species with a nest volume less than 5 litres). Read more about hive design and volume in Chapter 9.

TABLE 7-2 Average worker bee size and nest characteristics of the Australian species of stingless bees.

Scientific name	Worker length (mm)	Nest volume (litres)	Brood volume (litres)	Brood structure	External entrance tunnel	Common name
Tetragonula carbonaria	4	8	2	Regular comb	No	carbonaria
Tetragonula hockingsi	4.5	10	2	Semi-comb	No	hockingsi
Tetragonula davenporti	4	8	2	Semi-comb	No	davenporti
Tetragonula mellipes	4	4	0.5	Semi-comb	Usually	mellipes
Tetragonula clypearis	3.5	4	0.5	Cluster	Usually	clypearis
Tetragonula sapiens	4	4	0.2	Cluster	Usually	sapiens
Austroplebeia australis	4	5	0.5–1	Cluster	Variable	australis
Austroplebeia cassiae	4	8	1–2	Cluster	Usually	cassiae
Austroplebeia magna	4	2	0.5	Cluster	Usually	magna
Austroplebeia essingtoni	3.5	1	0.2	Cluster	Usually	essingtoni
Austroplebeia cincta	3.5	4	1	Concentric layers	Yes	cincta

Common names for Australian stingless bee species

Stingless bees are known collectively in Australia as sugarbag bees, bush bees or (rather nebulously) native bees. Generally, no common names exist for various species because they are so difficult to distinguish from each other. Efforts to find Aboriginal common names for species have not been fruitful.

With increasing knowledge of the distribution and characteristics of the various species, many people have sought unique common names for accurate reference to species. It is becoming customary to use the specific name without the genus as the common name (carbonaria, hockingsi, australis, etc.) (Table 7-2). This usage cannot cause confusion, as no Australian species share a specific name. We support this usage, although in this book we have stuck to scientific parlance, whereby a species name includes both genus and specific name in italics, for example, *Tetragonula carbonaria*, shortened to *T. carbonaria*. Genus names can also be used as a common name if you are referring collectively to all members of that genus, so the common name Austroplebeia can be used for the genus *Austroplebeia*.

Name changes of the two genera of Australian stingless bees

Until 1990, all Australian stingless bees were called *Trigona*. The authoritative book *Insects of Australia* (CSIRO 1991) classified all of these species as *Trigona*, but divided them into the subgenera *Plebeia* and *Tetragona*. This followed a system of classification proposed by the late, great, American bee biologist, Charles Michener.

In the 1960s, an alternative system of naming was proposed by the Brazilian bee expert Jesus Moure. Moure recognised many more genera than Michener. Michener responded that numerous names would be meaningless for everyone but a few specialists, so he combined 120 species from the Americas, Australia and Asia into the *Trigona* mega-genus. Moure's system tended not to be used as widely as Michener's. Moure died in 2010, but his spirit lives on because most of his taxonomic system is now widely accepted.

So how did the name changes transpire? First, Michener showed in a new analysis that the group of Australian stingless bees that were previously called genus *Trigona,* subgenus *Plebeia,* came out on an entirely different branch from the *Trigona*. They needed a new name to indicate their status. Moure had earlier recognised them and had proposed the name *Austroplebeia*, which Michener resurrected. Recent analyses have confirmed this change. In particular, the excellent work of Danish researcher Claus Rasmussen and colleagues has produced a convincing phylogeny of the world's stingless bees, based on DNA molecular sequence data, rather than morphological traits.

Rasmussen's phylogeny also shows that the remainder of the Australian stingless bees called *Trigona* (and their Asian relatives) are not related to American *Trigona*. The first *Trigona* named was a South American species, so that group got to keep the name according to the rules of naming animals. This meant that the Australian/Asian bees needed a new name. Like *Austroplebeia,* a name already existed; the name *Tetragonula* was earlier proposed by Moure. It turns out that Moure's genera closely align with the natural groups revealed in Rasmussen's system. Rasmussen's work validates the generic system of Moure, and justifies the use of the name *Tetragonula*.

Nobody likes it when plant and animal names change, but there are clear reasons for, and advantages to, making these changes. It does not make sense to call all these species *Trigona* when they are not closely related. It clouds our thinking about patterns. When groups are natural ones, then we can predict aspects of the biology of species based on what we know of others in the group. We can study the evolution of a trait (e.g. recruitment behaviour) in the context of knowing the evolutionary history of the species we are studying.

This genus is represented by six species in Australia. Many other species occur in Asia. In fact, *Tetragonula* is the single largest and most widespread stingless bee genus in the Indo-Malayan / Australasian regions, reported from India to the Solomon Islands.

Two *Tetragonula* species are found in both Asia and Australia (the two species from far north Queensland, *T. clypearis* and *T. sapiens*). Other species (*T. carbonaria*, *T. hockingsi*, *T. mellipes*, and *T. davenporti*) are endemic to Australia. These four endemic species are related, and fall under the banner of the *carbonaria* group. The members of the *carbonaria* group are similar in morphology (form and appearance), but their nest architecture differs. Furthermore, recent studies from Ben Oldroyd's lab on the DNA sequencing of the four members of the *carbonaria* group confirm that they are all distinct species that rarely interbreed to produce hybrids. Just a few cases of *T. hockingsi* males mating with *T. carbonaria* queens have been identified, as has a hybrid between *T. carbonaria* and *T. davenporti*.

It is not easy to separate the *Tetragonula* species and, usually, a combination of worker morphology and nest architecture is required to do so. One characteristic of interest in identifying the species is the hair on the side of the thorax. In *T. clypearis* and *T. sapiens*, the front section is densely haired while the middle section is shiny and sparsely haired. This distinguishes them from *T. carbonaria*, *T. hockingsi*, *T. mellipes* and *T. davenporti*, in which the whole thorax side is evenly covered with fine hair (Figure 7-10). Use of this and other characteristics requires a series of good specimens, experience, patience and a reasonable microscope. For a full description of how to identify species, see the following resources:

- *How to Recognise the Different Types of Australian Stingless Bees* by Anne Dollin — Booklet 4 in the Native Bees of Australia Series (www.aussiebee.com.au/booklet-4.html)

- the Australian Pollinators section of the PaDIL website, which contains fine images by Ken Walker (www.padil.gov.au)

- "Australian stingless bees of the genus *Trigona* (Hymenoptera: Apidae)", by A. Dollin, L. Dollin and S. Sakagami (1997), in *Invertebrate Taxonomy*, 11: 861–896.

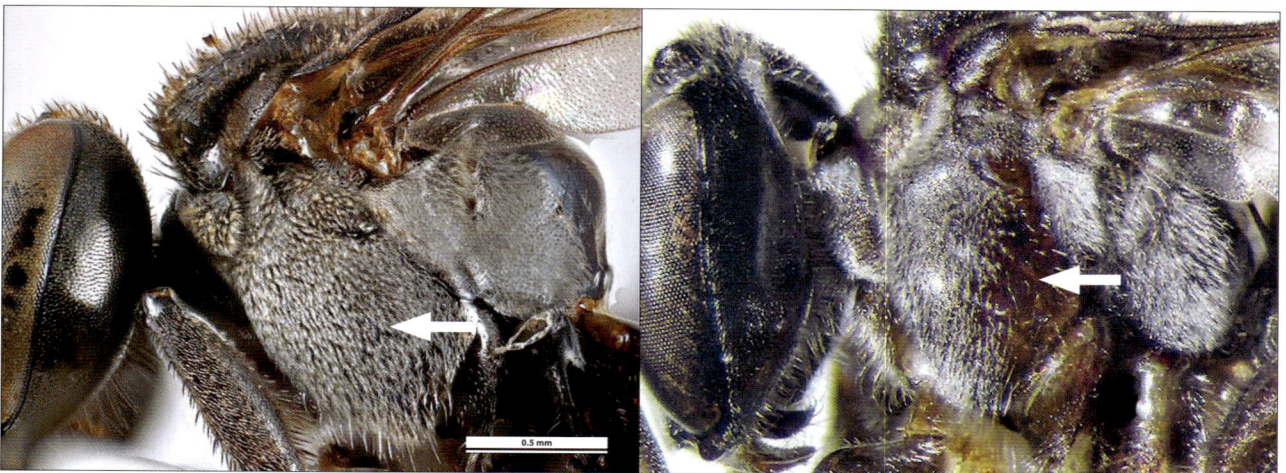

FIGURE 7-10 The covering of hairs on the side of the thorax is even in members of the *carbonaria* group of bees, as in *Tetragonula carbonaria* in the image on the left (shown by arrow). IMAGE **KEN WALKER**
But this covering is uneven in *Tetragonula clypearis* and *T. sapiens*, as in the sparse area in *T. sapiens* in the image on the right (shown by arrow). IMAGE **ANNE DOLLIN**

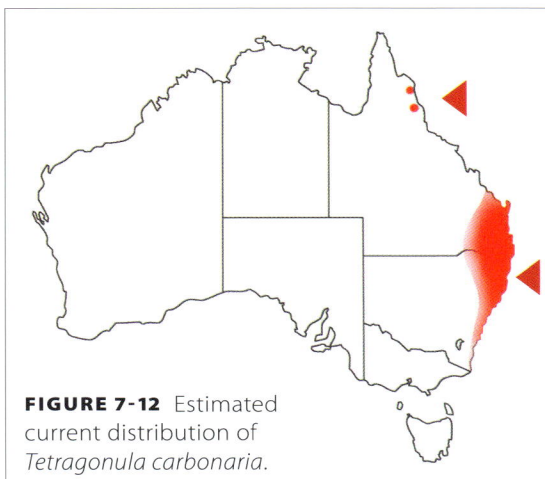

FIGURE 7-12 Estimated current distribution of *Tetragonula carbonaria*.

FIGURE 7-11 The advancing front of the brood comb of *Tetragonula carbonaria* is in the shape of a flat spiral that may be clockwise, anticlockwise, off-centre, double or multiple.

IMAGES **MIKAYLA LAMBERT** AND **GIORGIO VENTURIERI**

FIGURE 7-13 *Tetragonula carbonaria* forager.
IMAGE **TOBIAS SMITH**

7.3.1 *TETRAGONULA CARBONARIA*

This is the flag-bearer of stingless bees in Australia, being the most commonly kept species and better known and studied than any other. It is easily identified by the stunningly beautiful regular brood comb in the shape of a flat spiral (Figure 7-11).

Its distribution is subtropical, and covers an area of Australia from around Bundaberg in Queensland to the south coast of New South Wales (Figure 7-12). Its distribution to 36.4°S makes it the most temperate species of stingless bee in the world. It also occurs in high-altitude tropical areas of the Atherton Tableland and the Daintree. Unlike other species in Australia, its distribution coincides with relatively dense human populations; in fact, almost half of Australia's population lives in the natural range of this bee. Other reasons for its popularity are that it is very easy to keep and to propagate, and is a good pollinator and honey producer. Many sections of this book use *T. carbonaria* as the primary case species and then cover other species as variants.

Tetragonula carbonaria is the most commonly traded Australian stingless bee, and is regularly sent from beekeepers in Queensland to buyers in New South Wales. In cultivation, it is recommended for areas from Nowra north.

However, as you move inland from the coast and the moderating influence of the ocean, the climate becomes less favorable for this species. This is especially true where you gain altitude, with areas such as Canberra and Katoomba definitely not suitable. But an area like Camden in New South Wales, 50 km south-west of Sydney, 34°S, about 35 km from the coast and 100 m above sea level, is suitable. Areas where you can grow bananas perennially and they bear fruit may also be suitable to keep *T. carbonaria*. Areas with frosts may be suitable if winter days are warm. Remember, though, that while *T. carbonaria* can survive into the temperate zone, they are still really a subtropical species, so hives in cooler areas need to be carefully positioned and managed. *T. carbonaria* begin foraging activity above 18°C.

7.3.2 *TETRAGONULA HOCKINGSI*

This is also a very commonly kept species in Australia, but only in Queensland (Figure 7-14). It is more tropical in its distribution than *T. carbonaria*, and is only very rarely found south of the border (the few nests known in New South Wales are probably the result of human movement).

T. hockingsi workers are a little larger than *T. carbonaria*. This is most apparent when they bite you! Some regard them as more defensive than *T. carbonaria*, but both species are capable of biting strongly when their hives are opened.

T. hockingsi can form larger colonies than *T. carbonaria*, with more stored food and a larger quantity of brood (Table 7-2). Also, the brood of *T. hockingsi* is clearly different architecturally. The brood arrangement is a semi-comb as opposed to the comb of *T. carbonaria*. In a semi-comb, the brood cells are constructed in groups at the same height, but a new group of cells is built slightly above or below this height, disrupting the evenness (Figure 7-15).

Finally, the entrances of *T. hockingsi*, which usually lack resin deposits, can generally be distinguished from those of *T. carbonaria*, which are surrounded by such deposits (Figure 7-16). In north Queensland, *T. hockingsi* is known for collecting wet paint as a propolis substitute and for collecting sweat from human skin. *Tetragonula hockingsi* begin foraging activity above 20°C.

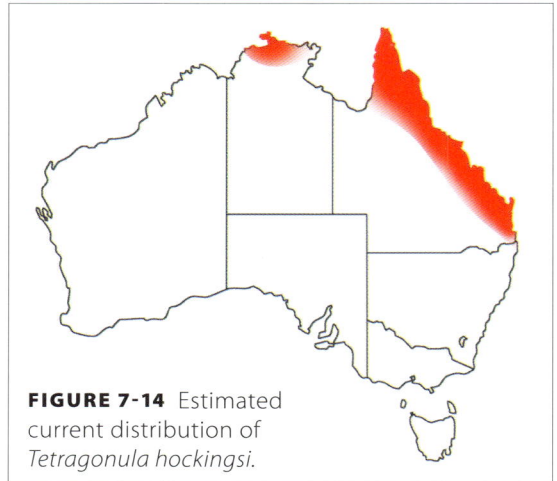

FIGURE 7-14 Estimated current distribution of *Tetragonula hockingsi*.

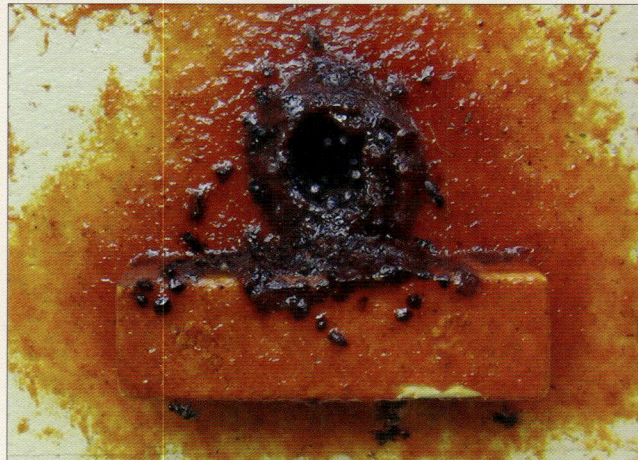

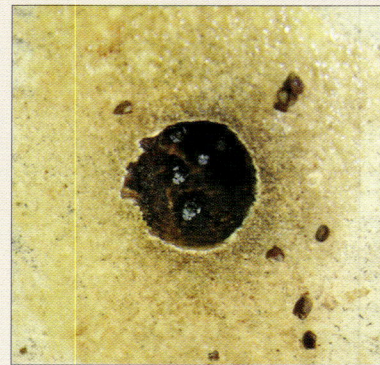

FIGURE 7-16 A comparison of the nest entrances of *Tetragonula carbonaria* (TOP PHOTO) and *T. hockingsi* (BOTTOM).

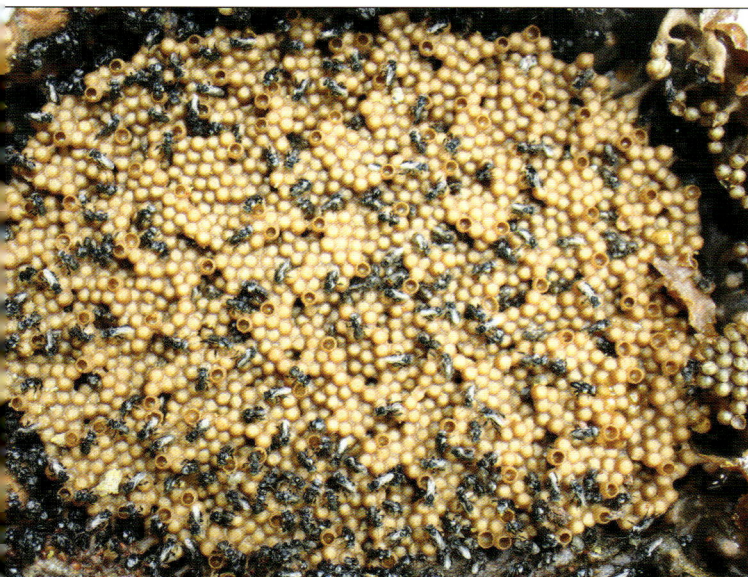

FIGURE 7-15 The advancing front of *Tetragonula hockingsi* showing the irregular semi-comb.

7.3.3 *TETRAGONULA MELLIPES*

This is the common *Tetragonula* of the north of the Northern Territory and Western Australia (Figure 7-17). It can be found nesting in small cavities inside trees and stone walls. It has a semi-comb arrangement of brood (Figure 7-18). The brood is also smaller, typically being less than half a litre compared with the average 2 litres of its relatives. Most *T. mellipes* nests have external entrance tunnels (Figure 7-19), while these are absent from the nests of *T. carbonaria* and *T. hockingsi*.

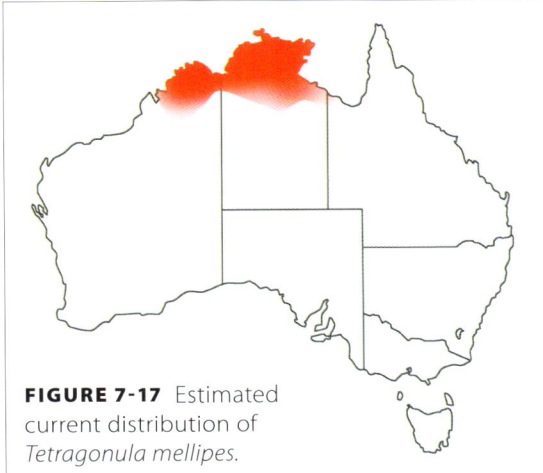

FIGURE 7-17 Estimated current distribution of *Tetragonula mellipes*.

(Further research is needed to confirm possible populations in western Queensland).

FIGURE 7-18 *Tetragonula mellipes* brood chamber removed from a log.
IMAGE **RUSSELL ZABEL**

FIGURE 7-19 Most *T. mellipes* nests have an external entrance tunnel.
IMAGE **AUNG SI**

7.3.4 *TETRAGONULA DAVENPORTI*

This species was named after Peter Davenport of the Gold Coast, a keeper of stingless bees who recognised it as a different species. Peter was later vindicated by genetic studies that proved his observations to be true.

T. davenporti is regarded as a cryptic species because it is so hard to distinguish from *T. hockingsi*. The brood is a semi-comb like that of *T. hockingsi* (Figure 7-20), but the nest is smaller and the bees are less defensive in behaviour. It is believed to be a rare species from a small area of southern Queensland (Figure 7-21), but its exact distribution and abundance are very difficult to determine because it can only be identified using sophisticated genetic tools.

FIGURE 7-20 The brood comb of *Tetragonula davenporti* resembles that of *T. hockingsi*.

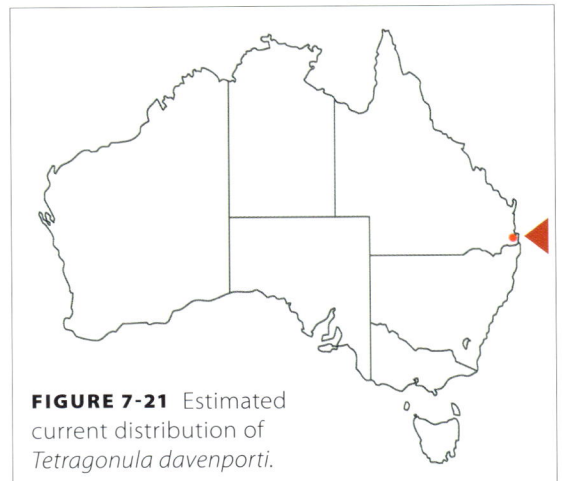

FIGURE 7-21 Estimated current distribution of *Tetragonula davenporti*.

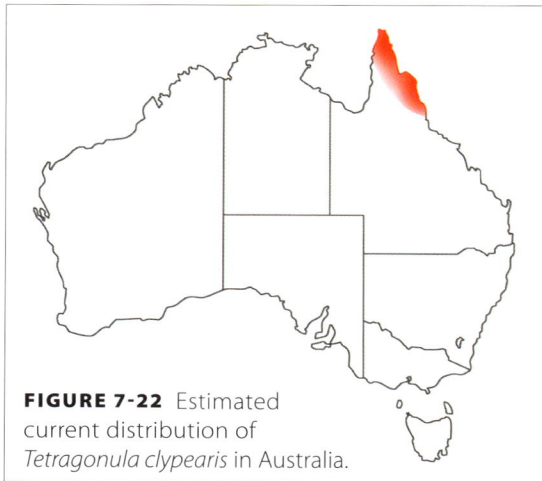

FIGURE 7-22 Estimated current distribution of *Tetragonula clypearis* in Australia.

FIGURE 7-23 *Tetragonula clypearis* workers flying into their nest; note the brown abdomens. IMAGE ANNE DOLLIN

FIGURE 7-24 *Tetragonula clypearis* tunnel in a house wall in far north Queensland. The nest is inside the conduit pipe. IMAGE ANNE DOLLIN

FIGURE 7-25 *Tetragonula clypearis* in a hive box. IMAGE ANNE DOLLIN

7.3.5 TETRAGONULA CLYPEARIS

Tetragonula clypearis is a widespread Indo-Pacific stingless bee, ranging from the Philippines to northern Australia. It belongs in the *iridipennis* group and so is closely related to the common Asian species *T. iridipennis* and *T. fuscobalteata*. This species is found in the high-rainfall areas of tropical far north Queensland (Figure 7-22). The workers look noticeably smaller than other *Tetragonula* species in Australia (Table 7-2, Figure 7-28). Another difference is that, in many colonies, the workers can have brown abdomens (Figure 7-23) rather than the black abdomens of mature adults of other species.

Some even smaller bees were collected from Cape York in 1993. Local Aboriginal people called them "mosquito bees". They looked like *T. clypearis*, but it was reported that they could be the minute Asian species *T. fuscobalteata*. On a safari in 1998, Anne and Les Dollin found that some Cape York *T. clypearis* nests contained workers just as small as the mosquito bees. A DNA study is needed to be sure, but it seems that these mosquito bees may be just a smaller variant of the species *T. clypearis*.

Nests of *T. clypearis* are very commonly found in the wall cavities of houses, particularly wooden walls. Nests often have an external entrance tube up to 15 mm long (Figure 7-24) but they can be up to 180 mm. This species can be kept very successfully in a wooden box about half the volume of that used for *T. carbonaria*

(Figure 7-25). (Read more about this in "Mini OATH hives" in Chapter 9.) Colonies can be successfully divided. Hives can be kept in close proximity to each other and do not seem to show the aggressive interactions of *T. carbonaria* and *T. hockingsi*.

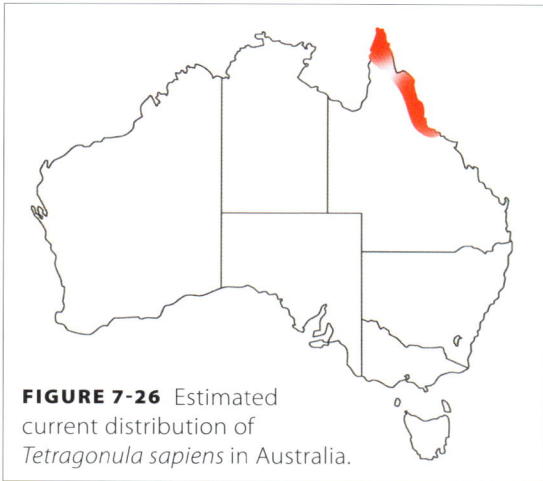

FIGURE 7-26 Estimated current distribution of *Tetragonula sapiens* in Australia.

FIGURE 7-27 A tunnel of *Tetragonula sapiens* with fallen leaves incorporated into it. IMAGE **ANNE DOLLIN**

7.3.6 *TETRAGONULA SAPIENS*

Tetragonula sapiens is another species that Australia shares with Asia. Its distribution is even wider than that of *T. clypearis*, ranging from the Philippines through New Guinea to the Solomon Islands. In Australia, this species is found in the high rainfall areas of tropical far north Queensland, similar to *T. clypearis* (Figure 7-26). It belongs to the *laeviceps* group of *Tetragonula* bees. Nests of *T. sapiens* are commonly found in the wall cavities of houses, particularly concrete block walls. It is reported to be a difficult species to successfully transfer into hive boxes. Nests typically have a short external entrance tube (Figure 7-27).

T. sapiens bees are noticeably larger than *T. clypearis* bees (Figure 7-28). The abdomens of *T. sapiens* workers are usually darker than those of *T. clypearis*. There is also a clear difference in the hair patterns on the hind legs of the males of these two species (Figure 7-29), though you would need to look at them under a microscope to see this. In addition, the internal architecture of the nest can be used to distinguish between *T. clypearis* and *T. sapiens*. While the arrangement of the brood cells is similar in these species, the nest contents differ. In *T. clypearis*, the honey and pollen pots are distant from the brood when the nest cavity is large (Figure 7-25). Although the pots are closer to the brood when the nest cavity is small, they never surround the brood as they do in the nests of *T. sapiens* (Figure 7-30).

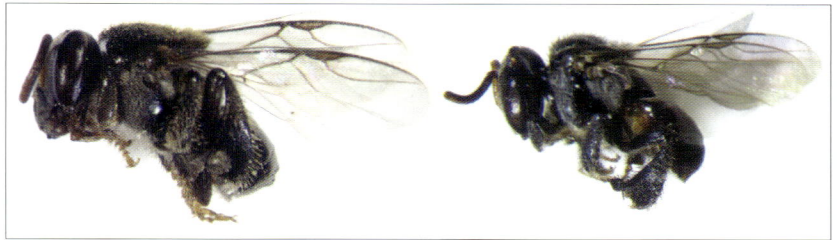

FIGURE 7-28 *Tetragonula sapiens* (LEFT) is noticeably larger than *T. clypearis* shown at the same magnification (RIGHT). IMAGE **ANNE DOLLIN**

FIGURE 7-29 Differences in the hair patterns on males' hind legs in *T. sapiens* (LEFT) and *T. clypearis* (RIGHT). In *T. sapiens*, the leg is relatively narrow and is evenly covered with sparse bristles. In *T. clypearis*, the leg is wider and has a mane of long feathery hairs on the back edge.

IMAGE **ANNE DOLLIN**

FIGURE 7-30 *Tetragonula sapiens* nest in a hive box. Note the proximity of the honey and pollen pots to the brood. IMAGE **ANNE DOLLIN**

7.4 More about *Austroplebeia*

This genus is represented by five species. It is known only from Australia, except for *A. cincta*, which also occurs in New Guinea. Some *Austroplebeia* are found in inland areas of lower rainfall and greater temperature variation than *Tetragonula* (Figure 7-3).

Most *Austroplebeia* species close their nest entry at night with a resinous lacy curtain that resembles a security grille (Figure 7-31). However, *A. cincta* colonies in Queensland sometimes leave their entrances open at night, or partially close them with a solid barrier of resin particles.

Inside the nest, the species of this genus show less use of propolis than *Tetragonula*. *Austroplebeia* use propolis to build their nest wall (batumen) around their nest but otherwise do not appear to use much propolis in other nest structures. The pots and other structures such as involucrum and connectives appear to be paler in colour (Figure 7-6). The honey pots are readily distinguished from the pollen pots because the thin pale walls allow the natural colour of yellow or amber to be seen (Figure 7-6).

It is usually straightforward to confirm that an individual bee belongs to the *Austroplebeia* genus by the presence of pale markings on the body and other characteristics such as the inside of the hind leg (Section 7.2). However, it has proven very difficult to separate the species of this genus. Anne Dollin has just completed a three-year study on the differences between *Austroplebeia* species, in collaboration with stingless bee expert Claus Rasmussen. This was based on data from more than 100 nests that Anne and her husband, Les, had studied throughout Australia on their native bee safaris. While the markings of these bees were their most obvious features, it turned out that it is far more accurate to distinguish the species by comparing the widths of their leg segments. For a full description of how to identify the *Austroplebeia* species, see the following resources:

- "Australian and New Guinean stingless bees of the genus *Austroplebeia* Moure (Hymenoptera: Apidae)–a revision", by A. Dollin, L. Dollin and C. Rasmussen (2015), in *Zootaxa*, Issue 4047 (1), pp 1–73.

FIGURE 7-31 Most *Austroplebeia* close their entrance at night with a veil of propolis. This is the entrance closure built by a boxed colony of *A. australis.* IMAGE **TOBIAS SMITH**

FIGURE 7-32 A hive of *Austroplebeia australis* with a removable top and clear observation window. IMAGE **CLAUDIA RASCHE**

- *Aussie Bee Online* article 25, "Meet the *Austroplebeia* species" which you can download from *Aussie Bee* website: http://www.aussiebee.com.au/abol-current.html

Observation windows work well in hives for *Austroplebeia*, as the bees do not cover the clear plastic with propolis the way that the *Tetragonula* species do (Figure 7-32). In addition, the brood is not usually enclosed in an involucrum and so is more easily seen. Finally, the queen (always a thrill to see!) boldly presents herself on the comb, rather than hiding within the comb as *Tetragonula* queens do.

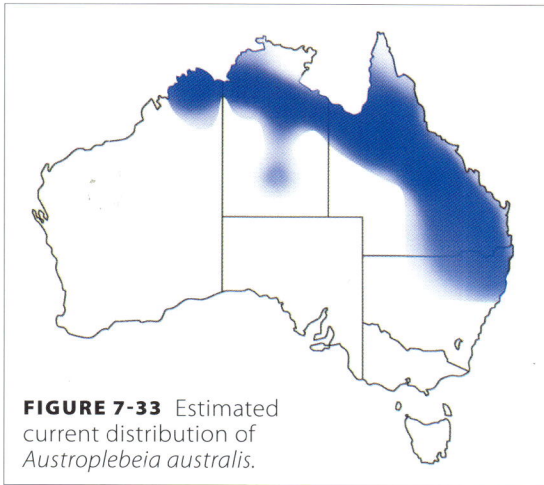

FIGURE 7-33 Estimated current distribution of *Austroplebeia australis*.

7.4.1 AUSTROPLEBEIA AUSTRALIS

In her revision of the *Austroplebeia* species, Anne Dollin found that the bees previously known as *A. australis, A. cockerelli, A. ornata, A. percincta* and *A. websteri* were, in fact, all the same species. All of these bees now should be called *A. australis* because this was the earliest name given to the group. This is a very widespread species, stretching from the north coast of New South Wales through Queensland and the Northern Territory across to the Kimberley in Western Australia (Figure 7-33).

In eastern areas, workers of this species usually have a row of four cream marks on the hind edge of their thorax. In central Australia, the markings on their face and thorax are brighter and more extensive. In western areas of Queensland, a particularly eye-catching variety has a glowing orange-coloured abdomen (Figure 7-34). However, Anne's analyses showed that all of these bees belong to the species *A. australis*.

Austroplebeia australis is abundant in drier areas of inland Queensland and New South Wales. It is very common, for example, on the Darling Downs west of Brisbane. Natural populations are also known in wetter coastal areas around Kempsey, Grafton, Gold Coast, and Bundaberg. In eastern areas, *A. australis* usually constructs a short tunnel as a nest entrance. However, in central areas of Australia this species often builds no entrance tunnel.

The brood cells of *Austroplebeia australis* are arranged in a cluster that resembles other *Austroplebeia* species (Figure 7-35). This species is commonly kept in Australia. Although it occurs naturally on the coast, it is much harder to keep there than in drier areas. Within its native range, this species survives very well and increases in weight in times of abundant flower. The honey of this species is thicker and less acidic than most other honeys produced by stingless bees.

FIGURE 7-35
A nest of *Austroplebeia australis*. The cluster brood is surrounded by a partial, single-layered involucrum and honey pots. Note that the honey pots are made of a thin layer of wax, not propolis, so are amber-coloured, not dark. IMAGE JEFF WILLMER

Many of the mysteries of this species have been revealed by the doctoral studies of Megan Halcroft. Megan estimates the brood populations of *A. australis* to be an average of 5,000 (range from 2,000 to 13,000) in natural nests, within its native range. The adult population is similar at approximately 4,000. Megan has shown that adult *A. australis* worker bees live for a surprisingly long time, up to 240 days. This is longer than any other stingless bee known.

Colonies of *A. australis* do not begin foraging activity until the temperature is above 20°C.

FIGURE 7-34 A western Queensland colour variety of *Austroplebeia australis* with a striking abdomen.
IMAGE DIANNE CLARKE

7.4.2 *AUSTROPLEBEIA CASSIAE*

Anne's recent study of the *Austroplebeia* species showed that the Queensland bees known as *A. symei* are the same as an older species called *A. cassiae*. Some bees from Mackay, Queensland, were given the name 'cassiae' by TDA Cockerell in 1910 because they were collected from *Cassia* flowers. The rules for naming animals say the older name must be used, so we now need to call these bees *Austroplebeia cassiae*. Generally, the worker bees are darker than those of *A. australis*. The hind edge of their thorax usually only has two ochre or cream spots, or no markings at all (Figure 7-36). Their face has thick white hair with at least one dull marking hidden underneath it.

FIGURE 7-37 Brood structure of *Austroplebeia cassiae* nests. Advancing front *(TOP)* IMAGE **TOM CARTER**
Detail of advancing front *(BOTTOM)* IMAGE **DON SMITH**

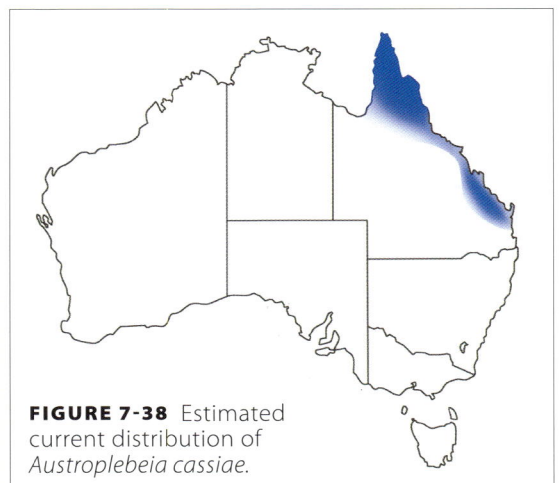

FIGURE 7-36 Side view of worker *Austroplebeia cassiae*.
IMAGE **ANNE DOLLIN**

The nests of *A. cassiae* are usually larger in size than those of *A. australis* (Figure 7-37) and they have more workers. They can be readily transferred into hives and propagated. This species is commonly kept in central Queensland, where it is naturally abundant (Figure 7-38). Tom Carter found that this species was very active on lychee flowers when hives were placed on farms.

FIGURE 7-38 Estimated current distribution of *Austroplebeia cassiae*.

7.4.3 *AUSTROPLEBEIA MAGNA*

In 1987, Anne and Les Dollin found a population of dark *Austroplebeia* bees in the Northern Territory. This species closely resembles *A. cassiae*. However, microscopic examination of the workers, males and queens revealed subtle differences, and showed that these bees were a new species. One distinguishing feature is that the worker bees have particularly wide basitarsus segments on their hind legs (Figure 7-39). So in their recent study, Anne and Les named this new species *Austroplebeia magna,* because "magna" means large. This species occurs in the top end of the Northern Territory (Figure 7-40).

In Arnhem Land, some worker bees are extremely dark, without any of the colour markings on the face or hind thorax that are seen in other *Austroplebeia* species. However, in the southern and eastern parts of their range, the workers have small colour markings, similar to *A. cassiae*. The males, as in all other *Austroplebeia* species except *A. cincta*, are much more colourful, with cream bands on the face, thorax and legs (Figure 7-39).

A. magna colonies usually nest inside small trees (Figure 7-41). Their nest entrance tunnels are often short but, like other *Austroplebeia* species, they may build quite long tunnels if the nest is being harassed by green ants (Figure 7-41).

BROAD BASITARSUS

FIGURE 7-39 A worker *(LEFT)* and male *(RIGHT)* of *Austroplebeia magna*, a newly described species from the Northern Territory. Note the broad basitarsus segment on the worker hind leg and the cream markings on the legs and thorax of the male. IMAGES **ANNE DOLLIN**

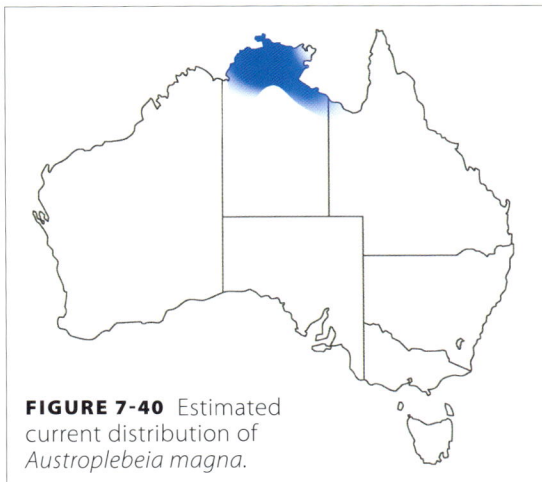

FIGURE 7-40 Estimated current distribution of *Austroplebeia magna.*

FIGURE 7-41 Nest structures and entrance tunnel of *Austroplebeia magna.* IMAGES **AUNG SI**

7.4.4 AUSTROPLEBEIA ESSINGTONI

Austroplebeia essingtoni is one of our smallest stingless bees. Both the nests (Chapter 7 opening page) and the workers (Figure 7-42) are diminutive. The species was named after Port Essington in Arnhem Land where it was first collected in 1840. This tiny bee occurs throughout northern areas of the Northern Territory and Western Australia (Figure 7-43). In some areas, the worker's face and thorax are very brightly marked, though in other areas the markings can be duller.

A. essingtoni usually builds a small round nest entrance tunnel (Figure 7-44). It nests in narrow twisty trees but will sometimes nest in cavities in rocky cliffs or stone walls. Preliminary attempts to establish this species in hives have been successful but propagation by division has not yet achieved much success.

FIGURE 7-44 The entrance tunnel of an *Austroplebeia essingtoni* colony that has nested inside a cavity in a rock cliff.
IMAGE **AUNG SI**

7.4.5 AUSTROPLEBEIA CINCTA

A brightly coloured bee from far north Queensland was determined by Anne Dollin to be *A. cincta*, a species previously thought to exist only in New Guinea. The worker has broad yellow bands on the top of its thorax, similar to many *A. essingtoni*. However, it also has bold yellow markings on its face and a yellow patch on the side of its thorax (Figure 7-45, Figure 7-46) that make it very easy to distinguish from other species. Surprisingly, unlike all other *Austroplebeia* species, the males are darker than the workers, lacking some of the thorax markings.

The brood chamber of this beautiful bee is also very distinctive because the new brood cells are waxed together into concentric layers (Figure 7-47). It also builds striking long entrance tunnels (Figure 7-48). In Queensland, one nest studied by Anne and Les had a tunnel that was 43 cm long! Green ants harass these nests and the bees keep adding sticky layers to the ends of their entrance tunnels to keep the ants at bay. In Australia, this species is only known from two locations in far north Queensland (Figure 7-49).

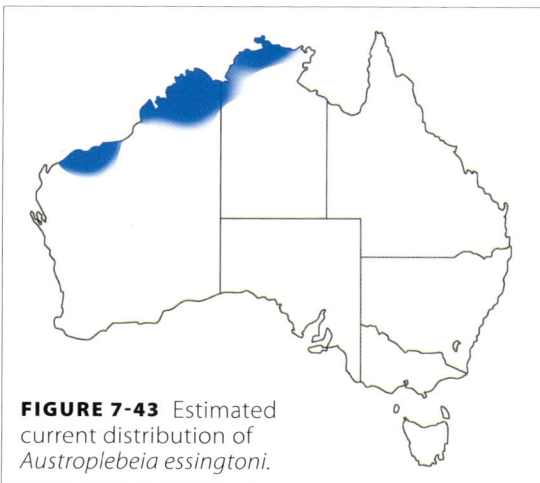

FIGURE 7-42 An *Austroplebeia essingtoni* worker from Broome in WA. Note the creamy yellow markings on the top of the thorax, not to be confused with the white hairs on the side of the thorax.

FIGURE 7-43 Estimated current distribution of *Austroplebeia essingtoni*.

FIGURE 7-45 Workers of *Austroplebeia cincta* at the nest entrance. Note the striking yellow markings on the face and thorax. IMAGE **ANNE DOLLIN**

YELLOW PATCH

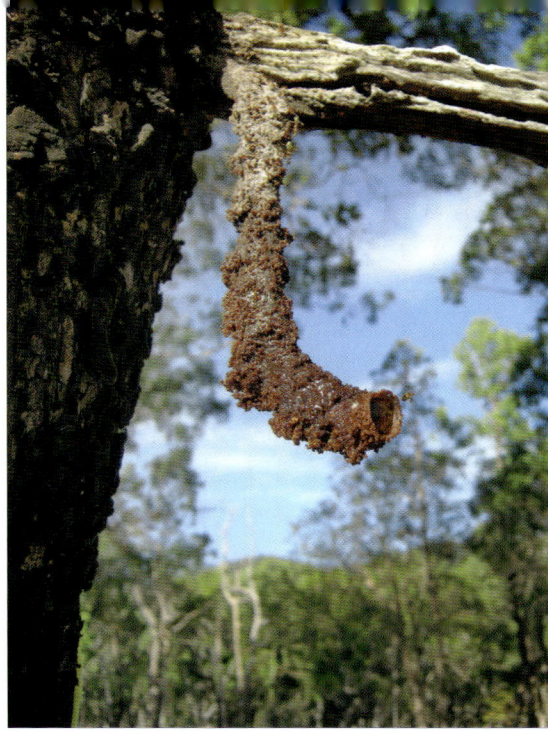

FIGURE 7-48 An entrance tunnel, 17 cm long, built by a nest of *Austroplebeia cincta*. IMAGE **ANNE DOLLIN**

FIGURE 7-46 The side of the thorax in an *Austroplebeia cincta* worker. Note the yellow patch beneath the wing that is an easy way to identify this species. No other *Austroplebeia* workers have this marking, except for rare, highly coloured *A. essingtoni* bees in Western Australia. IMAGE **ANNE DOLLIN**

FIGURE 7-47 Concentric layers of brood cells in the brood comb of an *Austroplebeia cincta* nest.
IMAGE **ANNE DOLLIN**

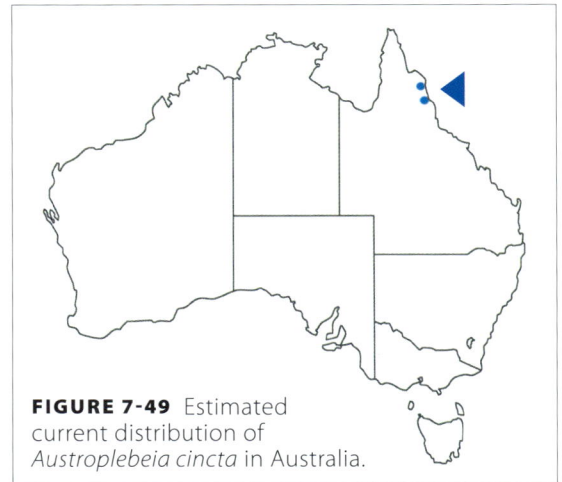

FIGURE 7-49 Estimated current distribution of *Austroplebeia cincta* in Australia.

FEATURE
Indigenous peoples and stingless bees

Shirley Ampetyane collecting sugarbag after cutting into a hollow tree that housed a stingless bee nest near Barrow Creek, Northern Territory, 2015. **IMAGE ALAN YEN**

Honey is a prized food in most human cultures. The ancient and intimate relationship between humans and honey bees in the Old World, including much of Europe, Africa and Asia, is well documented. It is less well known that the honey of stingless bees was a highly valued resource for indigenous peoples in warmer parts of the planet, including those in Australia and tropical America.

Indigenous Australians

Traditionally, the Indigenous people of northern Australia regarded the honey of stingless bees very highly, and it was eagerly sought. This honey was not a staple food, but played a significant role in rituals, art, mythology, religion and social life. Sharing and giving this special substance was important. It was also used as a medicine.

Propolis (referred to as "wax") was also valuable, being used in making implements such as spears and didgeridoo mouthpieces. Native bees and their hives were a popular subject for rock paintings.

Today, the word "sugarbag" is widely used by Indigenous people to describe native honey and the bees themselves. Sugarbag is still significant in Indigenous culture, and remains a common theme in Aboriginal art. It also appears as a clan totem in certain areas. Indigenous people recognise many different types of sugarbag, different parts of the hive, and stages of the bee life cycle.

In Queensland

Tom Petrie grew up with Indigenous Australians in Queensland in the 1800s and his faithful memories of his childhood experiences were later chronicled by his daughter, Constance Petrie. The first use that I can find of the word "sugarbag" appears in Constance's book, published in 1904, which documents Tom's memories from 1837.

In the book, Tom relates how Indigenous people found nests by looking for foraging bees or for the garbage pellets that dropped from nests to the ground. If the nest was high, they climbed the tree using a length of scrub vine looped around the trunk to support their ascent. The honey pots produced by the bees inside the hive were then cut out using a stone axe and dropped to those

waiting below. Some honey was eaten then, with the remainder carried back to camp in a "pikki" vessel made from bark. Indigenous people also used "braggain" (beaten and chewed fibre from the bark of the stinging tree *Dendrocnide moroides*) to mop up excess honey. Back at camp, the braggain was soaked in water to make a sweet drink and excess honey was later communally sucked from it.

Elizabeth McKenzie also grew up in an Aboriginal settlement in Queensland in the 1920s. She notes that, in addition to its value as food, honey was used to treat wounds and to stick wattle flowers to the body as a decoration for corroborees. She also describes a ceremony that tells the story of robbing a nest of native bees.

In the Kimberley

Reverend JRB (Bob) Love (a missionary who lived with the people of the Kimberley) and anthropologist Kim Akerman documented the role of honey and bees in the lives of local peoples in north-west Australia in the early and late 1900s respectively. They noted that stone axes were used to open the log containing the colony and fibre wads were used to collect spilled honey. If the log was too thick or hard to cut open, a small hole was cut, then a stick was inserted and agitated so that the honey flowed out into a bark bucket. Honey, pollen, adult bees, brood and wax were all eaten.

Honey was the sweetest food known to these people and they relished it (the honeypot ants of the central deserts do not occur in the coastal regions, where sugarbag bees abound). Honey was never cooked, but eaten raw. Occasionally, the honey was mixed with water and drunk. In the north-west, unlike the east, honey was collected by women. Along with meat and fat collected by men, honey was given as a gift to future parents-in-law or when visiting another group. These gifts were important in building social bonds and in economic exchanges. Honey was also presented at various ceremonies, including funerals. Dances and songs were performed to encourage an abundance of honey.

Wax was extracted by chewing or squeezing. It was then beaten and mixed with charcoal or ochre and rolled into balls or lengths, and used for purposes such as sealing containers, mounting stone axes and spears, making feather ornaments, and moulding ledges over cave paintings to protect them from water damage. References to wax refer to propolis (the mix of resin and wax that bees use to build nests). Today, many Kimberley communities still use sugarbag in these ways.

Following the bees back home

A well-known story about a technique used by Aboriginal people to find native bee nests can be found in "A Letter from Australia", published in the *New England Farmer* journal in 1861. A similar story is retold by J. K. Arthur in a book published in 1894. It comes from somewhere in the Upper Darling region, around 500 miles north-west of Sydney.

An Aboriginal man watches for a bee coming to drink at a pool. He has prepared himself with a tiny piece of down from the leaves of a plant. One end of the down has been rolled to a point and dipped into sticky resin from the broken stem of another plant. He then lies motionless over the edge of the pool with his mouth full of water until a bee comes to drink. Spitting the water from his mouth over the bee, he immobilises it, catches it, and allows it to dry. He then attaches the tuft of down to the bee's back and lets it go. The bee makes for its nest, but the tuft of cottony down serves the two-fold purpose of impeding its progress and showing a mark in the air for the man to follow. He pursues the bee to its nest in a hollow tree, then cuts the honey from the tree.

The future

Today, honey is still shared in Aboriginal communities, with some starting to keep stingless bees in wooden hives. For example, Russell and Janine Zabel have helped the Aurukun community in far north Queensland to develop a stingless bee industry that produces honey and propolis as major products. Other communities in northern Australia are pursuing similar ventures.

Native Australian bee hunters. From: Arthur, J. K., *Kangaroo and Kauri: sketches and anecdotes of Australia and New Zealand.* (London, Sampson Low, Marston & Company, 1894)

Lyndon Davis, of the Gubbi Gubbi people on Queensland's Sunshine Coast, holding one of his paintings representing a native stingless beehive in cross-section in a tree. The spiral in the centre represents the brood mass of *Tetragonula carbonaria* and the surrounding brown dots are the sugarbag honey pots.
IMAGE **CHRIS FULLER**

American native peoples

Indigenous people of the American tropics (from Mexico to Argentina) have long used the bounty produced by the many species of stingless bees that occur in this region.

Honey has always been the most important part of their harvest, providing a rich source of food and medicine. Wax and propolis also have a long history of use.

Bees of the genus *Melipona*, common in this part of the world, have large nests that can produce 5 kg of honey in a year. Traditionally, much honey was obtained from wild colonies, while some civilisations also domesticated bees. For example, in pre-Columbian times, the Mayans of Central America kept *Melipona beecheii* in hollow horizontal logs. The historian who accompanied Cortez to the Yucatan documented that thousands of hives were in use by the Indians, and honey and wax were plentiful. Ancient Mayan codices describe the practice of meliponiculture (keeping of stingless bee hives). Today, this practice, although rich in tradition, is on the verge of extinction.

However, another traditional form of keeping stingless bees still thrives in the mountains of eastern Mexico. The people here, of Totonaca origin, keep *Scaptotrigona mexicana* in two clay pots, the top one being inverted to join the bottom. The splitting method by which they propagate colonies is similar to that used today in Australia.

Throughout the American tropics, native peoples still use the honey of stingless bees for a wide range of medicinal uses, including the treatment of disorders of the skin and eyes and of the digestive, respiratory and female fertility systems. Pollen and propolis are also used in local therapies.

Traditional keeping of the stingless bee, *Scaptotrigona mexicana,* in clay pots in the State of Puebla, Mexico.

The next seven chapters of this book are dedicated to the noble science and art of keeping stingless bees.

Here you will learn how to build a hive box, establish a colony, divide your hive, extract honey, protect bees from threats, and lots of other fun and useful stuff.

Hold on for an amazing ride.

Transferring a stingless bee colony from a log into a box. IMAGE **GAIL BRUCE**

Part TWO

Keeping hives of stingless bees

Splitting a hive
of stingless bees.

8 Keeping stingless bees –getting started

Express yourself with a hive of stingless bees. Release the artist within.

Keeping stingless bees is gaining popularity in the warmer parts of Australia where these bees occur naturally.

Stingless bee hives **perform better in urban environments** with their rich floral resources, but can also do well on farms and in natural areas.

Hives need to be **protected from the hot sun**, but can be kept at moderately high densities and with honey bee hives.

To move a hive without bee losses, care and planning are needed.

Do not open your hive, only if splitting or extracting honey.

Stingless bees make great **"pets"**.

The technical word for the keeping of stingless bees is **meliponiculture**, coming from the scientific name of the stingless bees, the **Meliponini**. A collection of hives is called a **meliponary** and the person who keeps them is a **meliponist**. The honey bee (**Apini**) equivalents are **apiculture**, **apiary** and **apiarist**. In English, the common words are **beekeeping**, **bee yard** and **beekeeper**. The Latin-based terms meliponiculture, meliponary and meliponist have not taken off in Australia, so in this book I follow the widely used English terms.

While on the subject of definitions, now is a good time to revise some other pertinent terms. As we discussed in Chapter 4, a **colony** refers to a unit of bees cooperating to rear young. A **nest** refers to the physical structures, building materials and stored food of a bee colony in a natural location. A **hive** refers to a colony of bees, their stored food and building materials, and the artificial structure built by humans to house them. The word "hive" is also commonly used to refer to the artificial structure without the bees but, in this book, I use the expression **hive box** for that.

The keeping of bees is a form of animal husbandry, which is defined as the rearing and care of an animal for the purpose of production. It can equally be considered wildlife caring or the keeping of a pet. We will explore all of these perspectives in the following pages.

8.1 Where can you keep stingless bees in Australia?

Unfortunately, not all of the Australian continent is suitable for keeping stingless bees.

These bees occur naturally in Australia's warmer and wetter parts (Figure 8-1). I recommend that species only be kept in their natural range. You can read more about different species' natural ranges in Chapter 7 "The stingless bees of Australia".

Having said that, I know that some enthusiasts may want to keep species from other areas. In particular, people in Australia's southern parts often aspire to keep species that naturally occur farther north. You may be able to keep some species of stingless bees outside of their native range, but more care and maintenance will be required. Positioning and insulation of the hives will also be critical. Electrical warming of the hive under thermostatic control may be beneficial. Feeding the bees may also help them persist in marginal areas. However, it may never be possible to propagate hives in these areas. Harvesting honey from hives in these areas will also not be possible, because your bees will not be able to fly and forage in cool temperatures, so they will need their stored food to survive through winter.

Western Australian readers should be aware that it is illegal to import stingless bee colonies into Western Australia, but you can establish hives using locally native stingless bee species.

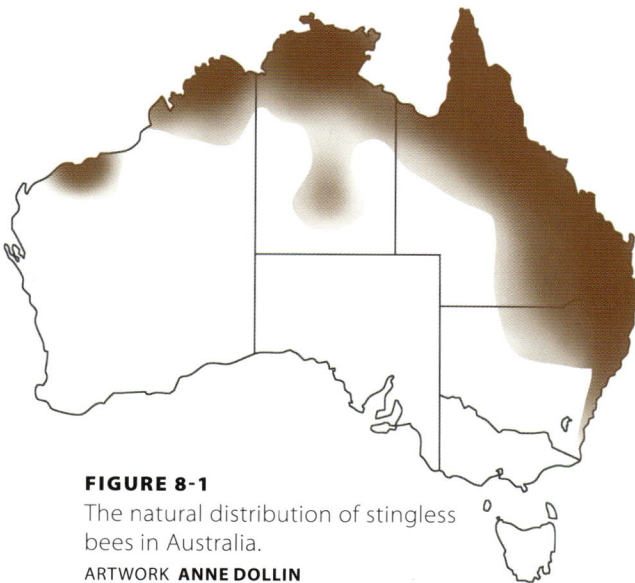

FIGURE 8-1

The natural distribution of stingless bees in Australia.

ARTWORK **ANNE DOLLIN**

The increased interest in stingless bees has stimulated trade and transport of colonies within Australia. It is easy to relocate them. (Read more about this in "Practical hints on transporting your hive", later in this chapter.) But it is not recommended, or even legal, to send colonies everywhere. Here are some points to consider before relocating colonies.

8.1.1 MOVING BEES TO WITHIN THEIR NATIVE RANGE

In general, it would appear safe to send colonies within the natural distribution of a particular species.

However, if their natural distribution is large, we should exercise caution when thinking about sending them long distances within that natural range. For example, *Tetragonula carbonaria* occurs from the Daintree and Atherton Tableland in north Queensland to southern New South Wales. Perhaps it is not a great idea to move a colony from one end of this range to the other and risk disrupting their genetic diversity and structure.

Moving colonies across great distances can create unnatural hybridisation events. Work from Ben Oldroyd's lab has detected three hybrid colonies that appear to have arisen because *T. hockingsi* colonies were moved from northern to southern Queensland where males mated with local *T. carbonaria* queens.

8.1.2 MOVING BEES TO OUTSIDE THEIR NATIVE RANGE WHERE THEY MAY SURVIVE IN THE WILD

This is not recommended. For example, *T. carbonaria* may survive in the wild in the south of Western Australia, but this area has its own bee fauna that could be adversely affected by the introduction of a stingless bee not native to the area. Don't upset the ecology by moving colonies there. In any case, it is illegal to send stingless bees to Western Australia from any other state.

8.1.3 MOVING BEES TO OUTSIDE THEIR NATIVE RANGE WHERE THEY ARE UNLIKELY TO SURVIVE IN THE WILD

There is probably little risk involved in sending species to an area where they are unlikely to survive in the wild. For example, sending a colony from north Queensland to Victoria to be kept by an enthusiast who uses heating and feeding to keep them alive is probably not harmful, because these bees are unlikely to found a wild colony and escape from culture.

8.2 Optimal environments for keeping hives

8.2.1 URBAN GARDEN — BEE PARADISE

We have discussed the climatic considerations that limit the keeping of various species of stingless bees. Now we need to answer the question: Within a particular climatic envelope, what are the best landscapes for stingless bees?

Australian species of stingless bees are found naturally, and are kept, in a diversity of environments, including inner-city urban areas, farmland and natural forests. Surprisingly, they perform best in urban environments. I worked with Sara Leonhardt, Benjamin Kaluza and other colleagues to test the performance of hives of *Tetragonula carbonaria* in various locations. We placed 16 hives in four sites of each of these three environment types and divided the hives when they reached a weight of 4 kg of contents. After four years, the initial 16 hives in each environment had greatly increased in number in urban areas but not in natural forests or on farms (Figure 8-2). Other studies have shown this to be the case also for honey bees and bumble bees.

This may sound counter-intuitive: surely native bees should perform best in natural environments? But many natural environments show extreme fluctuations in the availability of resources, while urban environments offer a more constant supply. In addition, resource density is greater in urban environments. Check this yourself: walk down a suburban street at any time of the year and observe the flowering plant density and compare this to a walk in the bush. The urban environment may be dominated by exotic species, but stingless bees will use any plants that provide suitable resources. In this sense, Australian stingless bees are adaptable and opportunistic species, able to thrive in habitats disturbed by humans.

Stingless bees performed poorly in natural forests and agricultural landscapes compared to the results recorded in urban environments. It is true that the farms chosen for this experiment were vast macadamia plantations, monocultures of a single species that flowers only for a short time. Farms that offer a diversity of bee forage could be more suitable. In fact, the hives did well on one farm with a relatively small proportion of high-quality riparian vegetation within flight range. But many agricultural land uses – such as grasslands for grazing, pine plantations, even some crops such as pineapples and custard apples – do not provide resources for bees. (Read more about using stingless bees in agricultural settings in Chapter 16 "Stingless bees for crop pollination".)

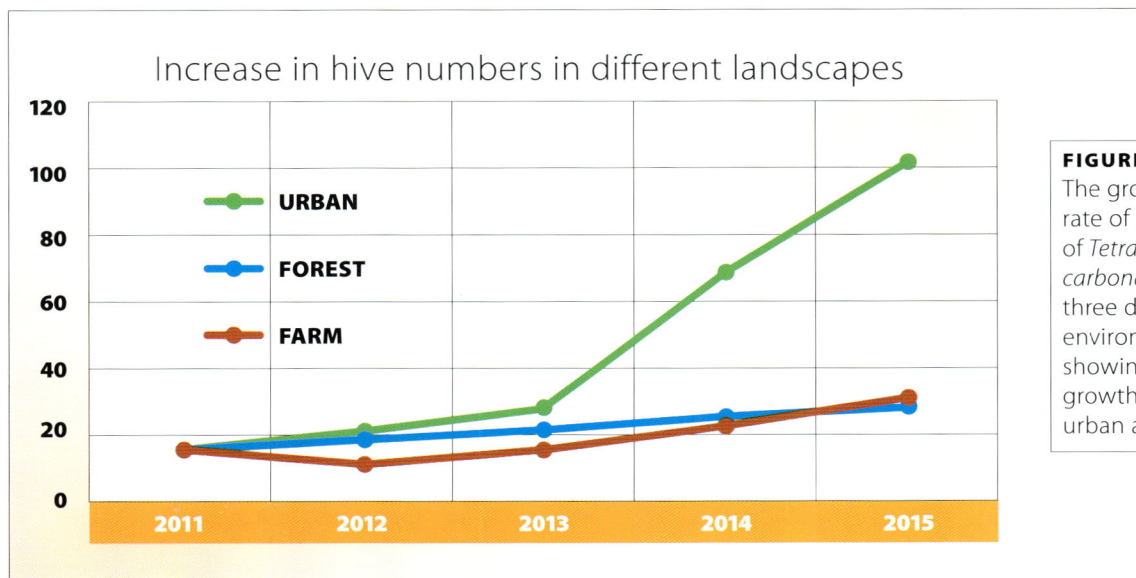

Increase in hive numbers in different landscapes

— URBAN
— FOREST
— FARM

FIGURE 8-2
The growth rate of hives of *Tetragonula carbonaria* in three different environments, showing rapid growth in urban areas.

8.2.2 PLANTING A GARDEN FOR STINGLESS BEES

Many people ask: What should I plant in my garden for native bees?

In answering this question, it's important to understand that even small stingless bees will forage over a distance of several hundred metres, so they can access flowers over a large area. Hence it is not so important what *you* plant, but what is available in their larger environment (Figure 5-9). But it is certainly the case that you can enjoy seeing bees more often in your garden if you plant bee-attractive flowers.

For ideas on plants to grow, see the "Attract Bees" PDF document available for free on the Valley Bees web page (mrccc.org.au/valley-bees/).

8.3 Positioning of hives

You should choose the best position for your hive before releasing your bees for the first time. Consider this carefully; you don't want to have to move your hive more often than necessary, so it is best to get the location right first time.

Honey bees are said to benefit from facing into the north-east quadrant, but I do not believe this is important for stingless bees. I have them facing in all directions on various properties and they do equally well. However, you do need to choose a suitable microclimate. Think about whether you would be comfortable sitting in your chosen position 24 hours a day, 365 days a year. Yes? Then the bees will probably be comfortable too. Remember that stingless bees do not need water (they get all they need from nectar), so there is no need to position the hive close to a water source.

On an urban house block, a covered outdoor area is usually a safe bet to position a hive. A position that receives morning sun in winter is optimal, because the warmth stimulates the bees to get active. Afternoon sun can cook a hive, especially in summer, so avoid exposed western positions. You can place the hive directly on the floor of a verandah or patio, but up on a stool, table, shelf, or other elevated position will help the bees take flight (Figure 8-3). Also remember that you will enjoy observing them coming and going, so position the entrance so that you can see it easily.

FIGURE 8-3 A hive installed on a shelf on a verandah, in an excellent position for both the bees and those wishing to enjoy their presence.

If you place your bees on a verandah or patio, be aware that they may drop garbage pellets close to the hive entrance. Normally, this material is transported and dumped far away from the nest but, in cooler weather, bees may drop it directly under the entrance. (Read more about this in "Removal of waste garbage pellets" in Chapter 4.) Although the pellets are not smelly or messy, it is best to position the hive so that they do not accumulate on your new deck!

How to best locate hives on a farm is discussed in Chapter 16.

8.3.1 DENSITIES OF COLONIES

In nature, stingless bees are found at quite high densities. Steve Maginnity rescued many wild hives from the widening of the Pacific Highway in northern New South Wales, and estimated a density of at least one hive per hectare. Russell Zabel has found 50+ colonies of *T. carbonaria* in a 50 ha block of protected land. Dean Haley found more than 100 natural hives in a patch of bushland in Brisbane, including 43 in a 50 ha section of that bushland. Allan Beil has recorded hundreds of nests of *T. carbonaria* and *Austroplebeia australis*. Megan Halcroft crunched Allan's data to show that the density of *A. australis* in an area of south-east Queensland averaged 0.6 nests/ha.

You can also keep stingless bee hives at moderately high densities, particularly in suburban areas. Home gardens are particularly rich sources of food for bees because so many ornamental plants flower heavily and offer abundant food for bees. Most suburban blocks will have spaces for at least four hives, maybe many more. When densities get very high, for example more than 50 hives on a hectare, competition will limit colony growth. In contrast, in areas that are poor for bees, problems of starvation could arise at hive densities much lower than this.

If you are keeping hives at high densities, be aware that fighting may become a greater problem when hives are closer. This may be due to fighting swarms (where one colony attacks another) or drift fighting (where bees accidentally enter the wrong hive and elicit strong defensive reactions). To avoid drift fighting, you should space out hives, place them close to landmarks, and mark the hives with shapes and colours to minimise the risk of foragers entering the wrong hive. (Read more about fighting swarms in Chapter 13.)

8.3.2 WHY DO MY BEES DIVE INTO MY POOL?

Pools and ponds can be an issue for bees. Bees use the light of the sky to navigate. Bees do not have circular canals like humans to tell them what is up and what is down. Instead, they perceive the sky with their eyes and maintain their position by keeping the sky up. If they leave a nest and see a large blue surface, they may think it is the sky and fly straight into it. I suggest that you find a position for your hive that is not directly above a pool.

8.4 Moving your hive

Colonies of stingless bees can be moved with no negative consequences. They quickly adapt to the new environment. Within hours of release in the new position, you may see the first bees coming home with pollen on their hind legs. But you do need to exercise some care and understand and apply some principles to ensure a successful move. The process of moving a stingless bee hive is similar to that of honey bees. The method you use depends on how far you wish to move the hive.

8.4.1 MOVING YOUR HIVE VERY SHORT DISTANCES (<1 M)

You can easily move a hive a short distance of no more than 1 m. Do this day or night, it does not matter. Some bees may initially return to the old position but, if the hive is less than 1 m away, they will perceive it and find the entrance. You can take short steps over successive days to move it distances of several metres. Taking short steps is a useful way of moving a hive into the shade in summer and out into morning sun in winter.

8.4.2 MOVING YOUR HIVE LONG DISTANCES (>1 KM)

It is also straightforward to move a hive long distances of 1 km or more. Pack the hive at night when the bees are all inside. Close all entrances, ventilation holes and drainage hole to stop the bees leaving the hive. After moving a hive to a new position, remove the plugs. Bees should immediately emerge (if it is daylight and above 18°C). They will do an orientation flight to learn the new position. In this orientation flight, they slowly back away from the hive, memorising the new position so they can find their way home after a foraging flight.

The distance of 1 km is significant because it is twice the foraging range of 500 m. If you move them less than 1 km, foraging bees may find themselves in familiar territory, their memory for the old position may prevail and they may return to that original position. But if the new position is 1 km or more, then this will not occur.

8.4.3 MOVING YOUR HIVE INTERMEDIATE DISTANCES (BETWEEN 1 M AND 1 KM)

Moving a hive an intermediate distance requires a little more effort and planning. If you move your hive a distance between 1 m and 1 km, some bees may forget the new location and return to the old location, never to be reunited with their hive. This will not kill the hive, but the foragers will be lost.

To move a hive an intermediate distance, a different strategy is required. To move from Point A to Point B, first move the hive from Point A to point C, which needs to be at least 1 km away from both Point A and Point B, and leave it there for at least three weeks. Then move it again to Point B.

Why do bees need to be moved at least 1 km for at least three weeks? We know the significance of 1 km as it relates to their natural foraging range. But what is special about this magic time of three weeks? This has to do with the length of life that an individual bee spends foraging. Foragers are adult workers at the end of their life. Their foraging stage is unlikely to extend beyond three weeks, so foragers with a memory of the old location will have been replaced at the end of three weeks. From this point on, there is no risk that foragers will return to the old location.

An alternative to moving the hive twice is to keep it closed for a period between the old and new locations. Dean Haley moves his hives around regularly but does not have an alternative site 1 km away to take them to. So he shuts them in a dark room for six days and then releases them in the new position. It seems that this is long enough for the foragers to forget the old position.

8.4.4 PRACTICAL HINTS ON TRANSPORTING YOUR HIVE

You can transport hives long distances over long durations in your car. With care, you can also send them in the mail or by courier to other areas. It should be safe for a hive to travel for up to five days if it is carefully treated. In particular, please observe the following.

- Pack the hive at night when all bees are inside.
- Close all entrances to stop the bees leaving the hive. Use something that can breathe, such as a piece of material, gauze or fly screen (Figure 9-10). Push the piece of material into the entrance loosely; do not pack it in. There may be more than one entrance and drain holes so make sure they are all plugged.
- Place the hive into a cardboard carton. Ensure that the top of the hive remains up in the carton.
- Arrange pieces of cardboard or other packing around the hive so it will not move in the carton.
- Tape the carton up well. Use a skewer to prick small holes into all sides of the carton. This will allow the carton to breathe but will not allow bees to escape if they find their way out of the hive.
- **Keep the hive cool**. Put your car air-conditioning on a low temperature. Do not leave the hive in the car if you have to leave it in the sun. Take it with you and leave it in a cool, shady place. Treat the hive like you would treat your pet dog.
- Avoid sending a hive if the weather is extremely hot.
- Avoid sending a hive if it is very full. Full hives have little air space inside and so are more susceptible to lack of oxygen than a hive that has a large volume of air space.
- Send a hive on a Monday to allow time for it to arrive at its destination before the weekend.
- If using a postal or courier service, choose a rapid service and write the following on top of the carton in large print:
  - Address and phone number of recipient
  - Where to leave if not home
  - Please handle with care
  - Keep this side up
  - Please keep cool

8.4.5 DO I NEED TO HANDLE THE HIVED COLONY GENTLY?

I strive to handle stingless bee hives carefully. I try to keep them in their original orientation. I lift them and place them gently. I do not vibrate the nest by scraping the box along the ground.

But I observe other successful beekeepers treating their bees with less apprehension! I have also seen how brutally some courier drivers throw boxes around when loading and unloading them. I have even seen a hive survive a fall of 4 m and then a long roll down a hill. It seems that the bees are able to withstand significant impacts. A clear exception is when the colony has recently been transferred to a hive and still has not fastened the hive structures together and to the walls of the hive. In this case, extra care certainly has to be taken.

8.5 Stingless bees as pets

They might not be warm and fuzzy, but they are sweet and buzzy. Stingless bees are intrinsically fascinating creatures, perhaps because their colonies, like those of honey bees, are perceived as a metaphor for human societies. Native stingless bees are gaining enormous popularity as pets (Figure 8-4). In fact, most people keep them for this reason (see the results from surveys of beekeepers below).

FIGURE 8-4 A stingless bee hive makes a great addition to backyards in the warmer parts of Australia *(LEFT)*. • Stingless bee hives are entirely safe for kindergartens and schools *(RIGHT)*.

As pets go, stingless bees have many benefits and few downsides. For example:

- You do not need a permit to keep them (only the hives of honey bees must be registered).

- They look after themselves.

- Colonies live for a very long time.

- They are easy to propagate to produce the ultimate gift or recoup the costs of purchasing them.

- There are no stings attached. They do not produce any venom, and are incapable of causing any serious allergies. In fact, they are entirely harmless, making them perfect as an educational resource in kindergartens and schools.

- They yield honey; it may not be produced in large quantities, but what is produced is delectable.

- You can contribute to the population of bees available for pollination.

- Native bees offer an opportunity to connect with the wonders of the natural world, even in an urban centre.

Urban keepers of stingless bees may also be able to do something about the decline of bees on farms. Stingless bees offer a viable alternative to honey bees for pollination of certain crops. The demand for alternative pollinators is expected to increase with continuing threats to honey bees. But populations need to be boosted in areas where clearing of trees has removed local remnant vegetation. Artificial propagation of hived stingless bees provides a means to boost populations. Local breeding on farms is often slow, because extensive agricultural ecosystems may not provide a continuous forage supply to support rapid colony growth. A viable strategy is to breed hives in urban areas and move them to agricultural environments. This is the reverse of most ecosystem services models, in which services are provided by natural areas and consumed in urban areas.

Many people working towards a more sustainable home find stingless bees a great addition because they pollinate the garden and yield a little honey. They often join other sustainability initiatives such as a compost bin, chickens and a worm farm. Those with a love of wildlife will add them to the native garden and frog pond in their backyard. People who live in apartments in cities and crave a little contact with nature can nurture them on their balcony along with their herb pots. Those who entertain visitors enjoy having them as talking points or items of curiosity. And curious they are, evoking feelings of fascination, contemplation and wonder, in ways that can be positively therapeutic. But be careful, bees are addictive! Your life may not be the same once you've been bitten by the native bee bug.

European honey bees are also wonderful pets. But there are some significant differences in keeping them compared to stingless bees. You should consider the pros and cons of meliponiculture vs apiculture before you embark on your beekeeping journey (Table 8-1).

TABLE 8-1 Comparison of keeping stingless bees and honey bees in Australia

	Stingless bees	Honey bees
Maintenance	Low	High
Need to register	No	Yes
Safety risk	Zero	Moderate
Cost of hive	High	Low
Cost of equipment	Low	High
Honey production (average kg/year)	1	50
Origin	Native to Australia	Introduced from Europe
Pollinators	Useful	Useful
Climatic zone	Tropics and subtropics	Everywhere but especially in temperate areas and subtropics

FIGURE 8-5 Hives of honey bees and stingless bees (and chooks!) co-existing peacefully at the Craig Farm. IMAGE **GLENBO CRAIG**

8.5.1 COMPATIBILITY OF STINGLESS BEES AND HONEY BEES

You can keep hives of honey bees and stingless bees together in the same yard (Figure 8-5). Generally, honey bees and stingless bees coexist happily (Figure 8-6). Very rarely, they may attempt to rob from each other. It is also possible that honey bees could deplete all the available resources, leaving little food for your stingless bees.

This may happen where densities of honey bees at a large apiary are very high. It is especially likely to happen in cooler areas, because honey bees work at lower temperatures and may deplete local resources each morning before it warms up enough for stingless bees to start foraging.

FIGURE 8-6 A honey bee and a stingless bee sharing a flower, non-aggressively. IMAGE **TOBIAS SMITH**

However, it is unlikely that there will be aggressive interactions between the two types of bees. It is also unlikely that they will spread diseases or pests to each other. None of the important diseases or pests of honey bees are contracted by stingless bees. The one exception is the small hive beetle, which may build up in the hives of honey bees and then attack stingless bees.

8.5.2 TRADING STINGLESS BEE HIVES

It has become quite common for keepers of stingless bees to sell hives. The vendor should provide a guarantee. A 12-month guarantee is important, because hives can continue to appear healthy externally for many months but be non-viable in the longer term. This is because workers can continue to collect pollen and other resources even in a colony that is queenless and doomed to fail. The guarantee can reasonably contain exclusions. For example, the guarantee may be voided if the hive experiences temperatures of greater than 42°C, receives direct sun after 10 am in summer, is kept outside the bees' natural geographical range, is opened, is split, or has honey extracted.

8.6 Snapshot of the state of keeping stingless bees in Australia

In 1998 and 2010, colleagues and I surveyed keepers of Australian native bees to learn more about their practices and their attitudes to beekeeping. Both surveys showed that most colonies were kept in suburban areas on the east coast of Australia, in an area centred on south-east Queensland. Three species accounted for the vast majority of hives kept: *Tetragonula carbonaria*, *T. hockingsi* and *Austroplebeia australis*. Most beekeepers used the Original Australian Trigona Hive (OATH) design or similar.

In the 12 years between the surveys, the number of beekeepers has more than doubled and the number of colonies has more than tripled (Figure 8-7). The sustainability of keeping stingless bees improved between the surveys, with a change from most beekeepers obtaining hives by transferring from natural locations to dividing existing hives.

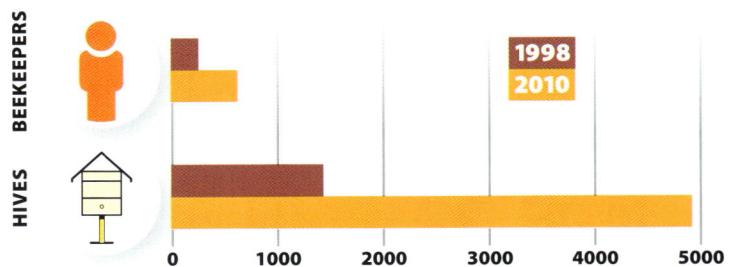

FIGURE 8-7 Increase in the number of beekeepers and stingless bee hives between the two surveys.

The second survey recorded high demand for Australian stingless bee hives and their honey, but supply was limited, with fewer than 250 beekeepers propagating hives in 2010. In both surveys, the top reason given for keeping hives was "enjoyment", suggesting that demand for hives is driven largely by the "pet bee" trade. Pollination services were provided by less than 4% of the major stakeholders within the industry.

Although Australia's stingless bee industry is growing at about 15% per year, it is still small compared to the honey bee industry, where 1,700 commercial keepers manage over 500,000 colonies. Table 8-2 summarises all the key findings of the surveys.

TABLE 8-2 Results of two surveys of keepers of stingless bees in Australia in 1998 and 2010.

	1998	2010
No. of beekeepers responding	298	635
No. of colonies kept	1425	4935
Most popular species kept		
• Tetragonula carbonaria	69%	62%
• Tetragonula hockingsi	20%	9%
• Austroplebeia australis	11%	23%
Nest locations		
• Suburban areas	56%	67%
• Near bush	24%	21%
• Rural areas	20%	13%
Reasons for keeping bees		
• Enjoyment	81%	78%
• Conservation	68%	67%
• Pollinate bushland	27%	29%
• Pollinate own crops	24%	24%
• Crop pollination services for others	–	1%
• Honey production	8%	11%
• Hive sales	5%	4%
• Education	2%	12%
• Research	2%	4%
State where beekeepers live		
• Queensland	71%	61%
• New South Wales	29%	38%
• Northern Territory	<1%	<1%
State where nests are kept		
• Queensland	91%	84%
• New South Wales	9%	16%
• Northern Territory	<1%	<1%
Honey production		
• Number of beekeepers	26	63
• Number of hives	542	1725
• Total honey production per year in kg	90	254
Bee colony propagation		
• Number of beekeepers only transferring nests	58	99
• Number of nests transferred	1240	5093
• Number of beekeepers only dividing colonies	17	139
• Number of nests involved	857	6328
• Number of beekeepers manipulating nests	119	238

9 Constructing hive boxes for stingless bees

Hives will provide a good home for a bee colony if you follow a few simple guidelines.

Hive volume is important to your bees' wellbeing. **Hive shape** makes little difference to bees, but a standard shape offers practical advantages.

Hives can be **built of various materials,** with pine wood being the most common and proven.

Hive designs come with **various features** such as split bars and observation windows.

The **Original Australian Trigona Hive** (OATH) is a recommended and standard design that facilitates colony propagation.

The **mini OATH** suits species with naturally smaller nests.

The **honey OATH** allow you to extract honey without damaging the brood.

Thinking inside the box!

In this chapter, I discuss the essential aspects of hive design and construction. For the past 30 years, keepers of Australian stingless bees have been striving to develop an optimal hive box design and have made great progress. I present the only design that has become widespread, the OATH. While there are other designs available (and I encourage you to look widely at various hive options), the OATH provides readers with a starting point that "ticks many boxes".

9.1 Fundamentals of hive box design

In developing hives for stingless bees, it makes sense to follow the principle of learning from the bees: that is, putting into practice what we know about bee biology.

For example, we have measured stingless bees' natural nest volume, so we build hives of a size that reflects this volume. We also know that they strive to maintain a constant nest temperature, but they are imperfect at this, so we provide excellent insulation. We know that they build a main entrance hole and also other vents, so we replicate those in our boxes. And we also know that they store honey in the parts of the nest that are most distant from the entrance, so we maximise the separation of the honey section from the entrance.

In addition to being great homes for their occupants, hives must work well for the keeper. If we combine these two goals – good for the bees, good for the keeper – we end up with the following important design elements for hive boxes:

- insulated, with appropriate entrances, correct volume, etc.: so they provide excellent housing
- comprised of two sections: so they can be divided for colony propagation
- light and compact: so they are easy and economical to transport
- durable and strong: so they last a long time, require little maintenance, and resist the elements, knocks, falls and attempts at entry
- simple: so that anyone can build one
- low cost: so they can be built economically.

Hive boxes that incorporate the above features will help us reach the goal of using Australian stingless bees on a large scale for pollination of crops, as pets and for honey production. If a box can be made commercially for $100 instead of $200, that's a saving of $10 million in the production of 100,000 hives, which is my rough estimate of the number of hives needed in Australia to contribute commercially to crop pollination.

9.1.1 HIVE VOLUME

Stingless bees disclose their preferred nest volume by their natural behaviour of building a nest wall when founding a new colony.

One of the first steps in designing hives was measuring this volume so it could be replicated. The process of transferring bees from their natural locations offered an opportunity to do this. This is best done in locations where the larger space is not limiting so that the occupied space is what the bees have selected. The typical volume in these situations is often around 1–10 litres, depending on the species (Table 7-2). For example, *Tetragonula carbonaria* is typically around 8 litres, *T. hockingsi* up to 10 litres and *Austroplebeia australis* around 5 litres. Building a hive of this volume ensures that the bees do not waste energy building a wall within the hive box to create their preferred volume. For comparison, the volume of a honey bee single box hive is about 40 litres.

9.1.2 HIVE SHAPE

The shape of the hive is not critical for the bees. Various shapes of artificial hives are used and are apparently equally successful. Stingless bees adapt to the variety of shapes found in nature and are equally adaptable in hives. I have observed a colony of *T. carbonaria* in a tree fern trunk that was 2 m long but only 60 mm in diameter. The bees had achieved a volume of nearly 6 litres in this long narrow space. So, if hive shape is not important to stingless bees, it makes sense to craft hive boxes in whatever shape is most convenient, practical and economical for their human keepers.

9.2 Building your box

9.2.1 BUILDING MATERIALS

Timber, PVC plastic, concrete, and other materials are being used successfully to construct stingless bee hives.

Insulation is of critical importance to protect against climatic extremes of heat and cold. (Read more about this in "Temperature regulation in social bee nests" in Chapter 4.) Plastic such as Nema Board is durable and a good insulator; however, it does not absorb liquid and is prone to condensation. So when splitting plastic hives it is imperative that you have adequate drainage to prevent bees drowning in spilled honey.

Timber is the most commonly used and proven material. Both hardwood, such as eucalypt, and softwood, such as pine, are suitable. Timber is an excellent insulator, but the denser the timber the poorer the insulation. Pine is light and so is a great insulator. It is also stable, economical, sustainably produced and easy to work with. It is only moderately durable, but lasts very well if kept painted and protected. After 30 years of service, most of my original hoop pine hives are still in good condition. Cypress pine is also popular because it is durable and cheap. If an old box deteriorates, it is not too onerous or risky to transfer the colony to a new one.

Timber thickness of 25 mm is often used, although it is common to see hives made of timber between 15 mm and 45 mm. The standard 25 mm wall thickness has proven adequate for keeping bees in moderate climes but, if you intend to keep them at the limits of their range, thicker wood will help.

Hardwood is more durable and does not have to be painted (Figure 9-1) but it has disadvantages relative to pine. To achieve the same insulating properties as pine of 25 mm thickness, you might need hardwood that is 40 mm, which will be expensive and heavy. A hoop pine box of 25 mm thickness weighs around 4 kg, but a hardwood box of the same thickness weighs around 8 kg, and a 40 mm hardwood box may weigh around 12 kg! Hardwood is also hard to cut, drill and plane, increasing the time and cost of production. In summary, hardwood makes good boxes if you are prepared to spend extra time, money and effort to make them and you do not expect

to move them often. You may also be able to find timbers that are both light and durable. For example, the softwood western red cedar is a fine material, but is expensive.

Timber must be well seasoned or it shrinks and warps, creating gaps. Bees will respond to gaps by filling them with propolis, making it difficult to separate the sections of the hive to divide it.

9.2.2 BUILDING TIPS

The joining surfaces where sections come together must be flat and not reveal gaps more than 0.5 mm. Gaps this small can be sealed very quickly by the bees, typically in around 24 hours. This gives protection against natural enemies attempting to gain entry through these gaps. Very importantly, minimising gaps contributes to insulation by reducing channelling of extreme temperatures into the box cavity.

Tight joints create access problems when you want to use a hive tool to separate hive sections. Use a wood plane to make a bevel on at least one corner of each surface where the sections meet so that you can gain entry with a hive tool (Figure 9-2). These surfaces (but not the inside of the box) can also be painted. The joints of each hive section need to be fastened with nails or screws and glued to eliminate gaps. Rebate joints are stronger than butt joints. If you are inserting any internal structures (such as brood excluders or split bars), be sure to rebate them, so that they do not prevent the joining surfaces of the sections coming together tightly.

FIGURE 9-1
A hardwood hive looks good, is very durable and does not need painting, but is heavy and hard to construct.

FIGURE 9-2
A bevelled corner of hive between the sections, for inserting the hive tool.

9.2.3 PAINTING THE HIVE

Paint the outside of your hive with three coats of exterior acrylic paint. Acrylic paint dries fast and releases little in the way of toxic fumes. Light colours offer superior heat reflection and longevity. White is the ultimate for these reasons but, if you prefer something else, make it a light colour. Paints with insulating properties are available and may help in extreme climates. To extend the life of hives, apply a coat of timber preservative, such as copper naphthenate, to the outside only of the box and allow it to age and off-gas for four weeks before painting.

You can re-paint the box either by touching up damaged areas or a complete re-paint. When re-painting, avoid painting the area within 50 mm of the entrance and ventilation hole, and paint at night or in winter when the bees are not active.

9.3 Hive elements and accessories

9.3.1 SPLIT BARS

Split bars are built-in structures located at the plane where the hive is divided (Figure 9-3, Figure 9-4, Figure 9-5). The bars prevent slumping of the nest in the top half after division. They also facilitate the colony division by helping to create a clean and even split between the two hive sections. Fewer honey and pollen pots are damaged when split bars are in place. There are numerous variants of the basic split bar idea, including side rails and split plates (Figure 9-6). Vertical mesh helps with slump but does not facilitate a clean split between the two hive sections (Figure 9-7).

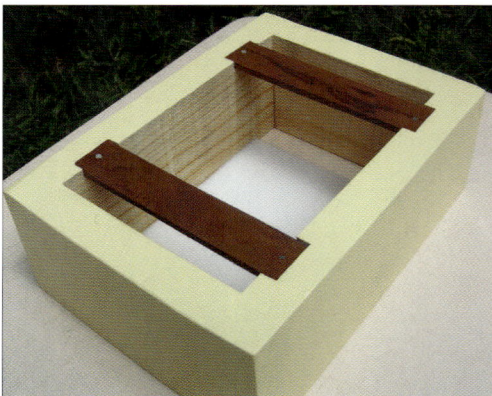

FIGURE 9-3 Split bars in the bottom of a top section (here turned upside down), can be made of wood (as in this case), metal or plastic.

FIGURE 9-4 Split bars (here made of galvanised metal strapping) can help to support the nest contents following the transfer into a box.
IMAGE **CLAUDIA RASCHE**

FIGURE 9-5 Split bars in an opened hive containing a thriving colony of *Tetragonula hockingsi,* showing the value of split bars when separating the two halves of the hive.

FIGURE 9-6 Chris Fuller hive with splitting plates separating the top and bottom sections.
IMAGE **CHRIS FULLER**

FIGURE 9-7 Peter Davenport hive with wire mesh in top section to reduce the risk of the top section's contents slumping after division.

9.3.2 OBSERVATION WINDOWS

A clear observation window for internal inspection provides a useful addition to a hive. It is best mounted on the top and covered by a removable lid to keep the hive dark and insulated when it is not in use (Figure 9-8). Observation windows are useful for managing colonies. The view through the panel gives an indication of hive strength. Internal inspection can help you to detect nest malaise early, allowing you to intervene if necessary or, at the very least, prevent the emergence of adult pests and their spread to other hives. Windows may also be useful for research and are an attractive addition to hives used for demonstration and education. (See a design for a hive with an observation window in "The OATH (Original Australian Trigona Hive)" later in this chapter.)

FIGURE 9-9 A false floor of wire mesh in the bottom section of a hive.

9.3.3 WIRE MESH FALSE FLOOR

A false floor made of wire mesh (Figure 9-9) can be helpful in a hive that you are using for a transfer of a colony into a box for the first time. In this operation, one expects that there will be spilled honey. (When dividing a colony, there is usually very little spilled honey and so the floor is not required.) Elevate the wire mesh a few millimetres above the wooden floor to create a space underneath for the honey to drain and to give access for the bees to repair the pots. Drill a hole through the basebaord to allow the honey to escape.

9.3.4 FEET

Give the hive short metal feet. Use screws or nails that extend about 3mm above the bottom surface of the box. Then, if the hive is placed on a flat surface, the feet will hold it above the surface and allow it to dry if water enters this space.

FIGURE 9-8

A hive with a clear observation window on top. The removable lid has been removed and placed beside the hive.

IMAGE **TOBIAS SMITH**

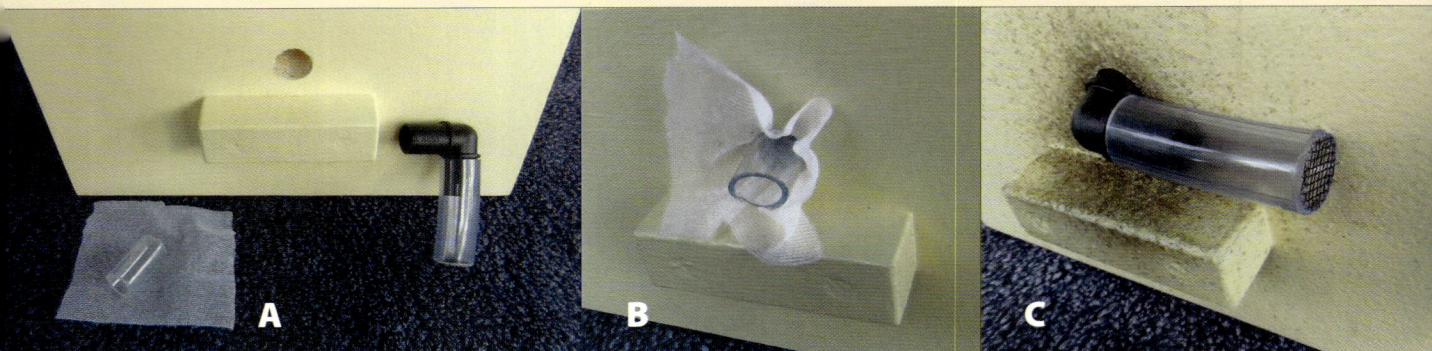

9.3.5 ENTRANCES AND CLOSURES

Stingless bees naturally build a hole that is approximately circular. This can be observed when they reduce the size of an existing hole (Figure 4-12). When building boxes, circular holes are also the easiest to make by drilling, and the easiest to seal with a plug. The diameter is important: too small and it may cause congestion among bees leaving and arriving, or may not provide adequate ventilation. Too big and it may take some effort for the bees to reduce the size and, in that time, natural enemies or extreme air temperatures may have penetrated into the nest. I use a 13 mm hole in normal hives and a 10 mm hole in mini-hives. Landing boards help those incoming foraging bees that do not get a grip the first time (Figure 9-10).

There may be an advantage to giving your bees extra protection at the hive entrance, especially for a period after a hive division. The bees will build a tube behind the entrance for defence but, in the meantime, they are vulnerable. In Brazil, where they have many problems with an aggressive phorid fly pest, beekeepers incorporate an artificial elongated entrance into their hives. This can be made by drilling a series of tubes into the front panel of the box behind the entrance. But most Australian beekeepers take other measures. (Read more about this in "Enhancing defences against pests that enter hives" in Chapter 14.)

Most beekeepers provide a second hole that the bees use primarily for ventilation (Figure 9-11). I drill a ventilation hole of 6 mm diameter at the back of the box, towards the top. The position of this second ventilation hole does not seem to be critically important, but a point that is distant from the main entrance would, in theory, allow better circulation of air.

FIGURE 9-10 Entrance and closures. Circular holes are easy to seal with a piece of tube of slightly smaller diameter. Landing boards help the incoming foragers.

A An entrance hole with landing board and two types of closures.
B A length of tube with a piece of gauze, which makes a perfectly fine closure.
C For a more robust structure, you can construct a closure with an irrigation elbow, length of tube and piece of metal gauze. This closure allows extra ventilation for longer periods.

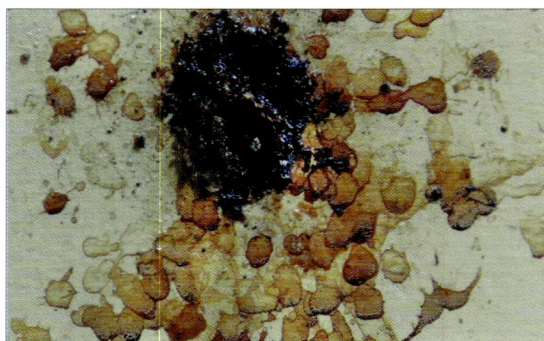

FIGURE 9-11 A ventilation hole doing its job. A guard bee is present, propolis structures protect the entrance, and drops of dried condensation show that the hole has been used to remove excess water.

9.3.6 HIVE MOUNTS AND STANDS

You will extend the life of your hive by protecting it from threats from below, particularly termites and rot-inducing dampness. Raising your hive above the ground will also help the bees to launch themselves into flight. Figure 9-12 shows options for mounting a hive.

One option is to mount a bracket on a wall and place your hive on a shelf (Figure 9-12 A and B). You can use a small screw to secure the hive to the shelf, either from below or through the side brace.

You can build a concrete block stand with two concrete blocks and a 300 mm square paver. Use single, double or, ideally, three-quarter blocks (Figure 9-12 C). Dig a hole and partially bury the bottom block. Fill the block cavity with soil to stabilise it. Use a concrete foundation for greater stability. Place the second block on top, but turn it 90° so the cavity is horizontal. Place the paver on top for a broader and more secure base for the hive. For added strength, secure the top block, bottom block and paver with an adhesive such as "Liquid Nails Landscape". Allow it to dry, then place the hive on top. Fasten the hive and roof to the base using tie-down straps or a Zabel's metal mini-emlock fastener. The horizontal top block provides an aperture for the strap to pass through.

FIGURE 9-12
A Shelf without the hive in place.
B Hive mounted on a shelf on a verandah.
C A simple concrete block stand.
D A hive mounted on a Y-picket.
E A durable and stable galvanised iron four-legged hive stand.
F A secure stand and cage for public places.

IMAGE C **CLAUDIA RASCHE**
IMAGE E **GLENBO CRAIG**

You can also mount the hive on a metal picket (Figure 9-12 D). Drive the picket into the ground. Attach a mounting bracket to the side of the hive. The mount can be made from a 100-150 mm section of 50 mm diameter PVC pipe. Alternatively, a proprietary picket rail mount is available at hardware stores. Place the hive on the picket using the mount. However, a disadvantage of these mounts is that you need to drill into the side of the box. The hole may rot, causing the screw to fall out, resulting in the hive falling to the ground.

Another alternative is to mount the hive on a stand constructed from welded lengths of metal (Figure 9-12 E). Galvanise the stand to prevent rust. Drive the stand into the ground with a mallet. Place the hive on top. Fasten the hive and roof to the base using the metal strap.

Use a sturdy steel mesh cage on a post in positions where security is a concern (Figure 9-12 F). The cage can include a roof and a lockable door. It will need to be concreted into the ground. Use a padlock to lock the door.

9.3.7 METAL ROOFS

Just like a wooden house, a wooden hive exposed to the elements needs a protective roof. An effective metal roof can be made from standard ridge capping used in roofing (Figure 9-13). Cut it to a length that extends over the front and back of the hive. Bend the side edges so they project inwards. Attach it by flexing it open and releasing it onto the top of the hive so that it clamps on.

9.3.8 POLYSTYRENE FOAM COVERS

A polystyrene foam cover (Figure 9-14) provides extra insulation against extreme cold or heat, which may be experienced in areas further from the coast. Peter Clarke recommends these covers for the Sydney region. You can keep this cover on permanently (as Peter suggests) or you can use it intermittently. Although the cover can keep out the cold on a frosty night, it will also prevent the hive warming up again the next morning. Likewise, it can keep out extreme midday heat but does not let the absorbed heat dissipate at night. An effective strategy for a hive located close to home is to put the cover on at night when it's cold and remove it the next day. Or, on a hot day, cover the hive in the

FIGURE 9-13 A hive with a protective metal roof.

morning and take it off at night. A disadvantage of polystyrene foam covers is that they harbour pests, such as spiders and other predators. Also polystyrene foam covers are voluminous, an important factor if you ever need to transport numbers of hives.

FIGURE 9-14 Hives on a stand with polystyrene foam covers.

9.4 The OATH (Original Australian Trigona Hive)

The OATH is the workhorse of Australia's stingless bee industry. The majority of beekeepers use it, or a variant of it, as the surveys discussed in Section 8.6 confirm. The OATH has proven ideal in many ways. It is simple to construct, so it can be built by anyone with average skills and modest tools. It is cheap, because the labour involved is minimal and the materials are economical. Its design also incorporates a means of splitting the hive, which has boosted its popularity and the number of new stingless hives being created. The OATH has propelled the propagation of countless thousands of new hives. The standard footprint has the advantage that it allows hives from various sources to be interchangeable.

How did the OATH get its name? At the time of its development in the 1980s, all stingless bee species in Australia were classified in the genus *Trigona*; the bee names have changed but the OATH acronym has stuck.

In addition to its suitability for colony division, the other significant aspect of the OATH is its shape and size, in particular the external footprint dimensions of 280 x 200 mm (Figure 9-15). A building material of 25 mm thickness is commonly selected for the OATH. This gives internal dimensions of 230 x 150 mm. This width neatly supports the brood chamber, including the involucrum, the natural width of which, when not size-limited, is around 150 mm in *Tetragonula carbonaria* and *T. hockingsi*. The extra space at each end provides appropriate food storage space and space to mount split bars to ensure they do not interfere with the brood.

9.4.1 THE STANDARD OATH HIVE

The standard OATH is made of two sections: top and bottom. The height of each section is typically around 90 mm, which gives an internal volume of 6 litres. This height improves the chances of the brood being present in relatively equal proportions when the two halves are separated. If the height is too large (e.g. greater than 100 mm), then the brood may confine itself to one half, reducing the probability of a successful colony division.

The box is perforated by two holes: a round entrance hole 13 mm in diameter and a ventilation hole 7 mm in diameter.

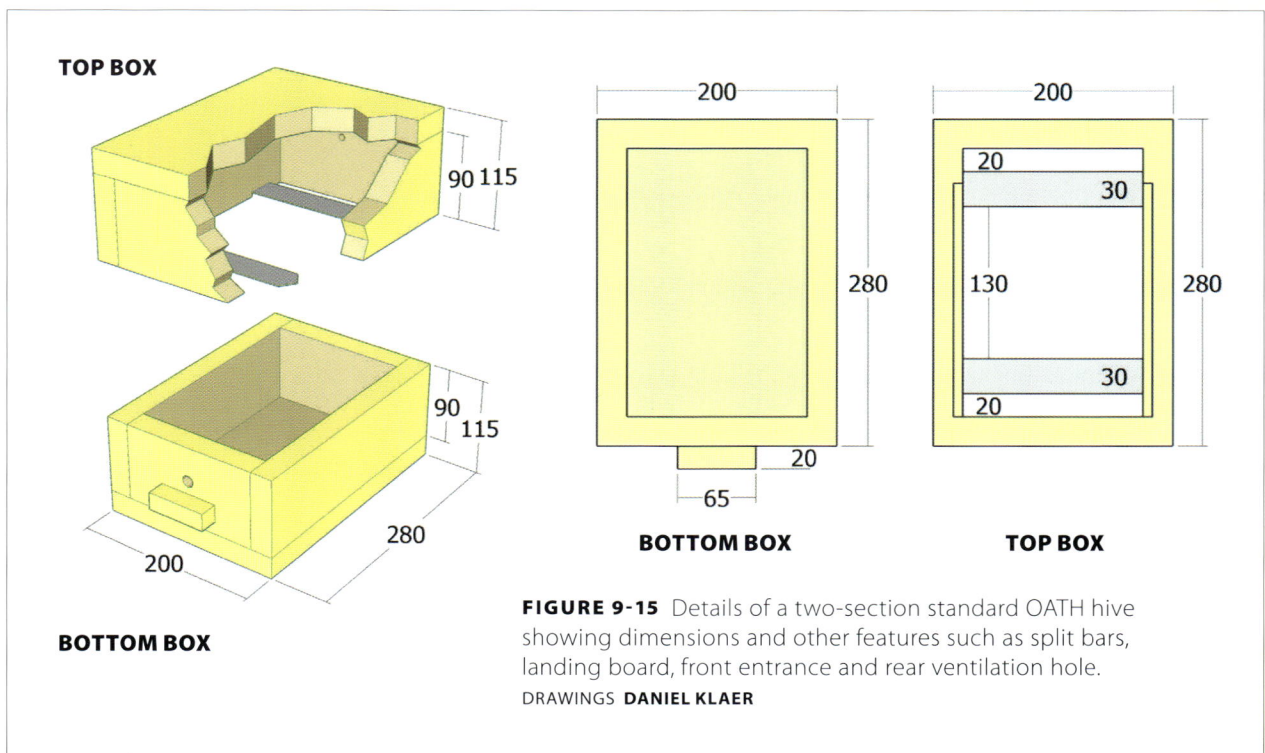

FIGURE 9-15 Details of a two-section standard OATH hive showing dimensions and other features such as split bars, landing board, front entrance and rear ventilation hole.
DRAWINGS **DANIEL KLAER**

9.4.2 THE OATH OBSERVATION HIVE

The OATH is compatible with a clear observation window for internal inspection. The window is fitted on the top and covered by a removable lid to keep the hive dark and insulated when the window is not in use (Figure 9-8, Figure 9-16). Care must be taken to ensure the window and lid do not allow entry of light, weather or natural enemies. Perspex and glass are popular choices for window material. However, thinner and more flexible clear materials – such as the acetate sheets once used for overhead projectors – have the advantages that they are cheap, readily available, easy to cut and easy to change. The ability to replace the material is particularly useful. Stingless bees tend to cover the clear material with propolis. In particular, *Tetragonula* bees may quickly cover observation windows, *Austroplebeia* less so. One way to keep these windows clear is to replace the clear material regularly.

9.4.3 HONEY OATH HIVES

The Honey OATH has an extra section on top, a **honey box**, **honey super** or **honey section** (Figure 9-17, Figure 9-18), which allows the removal of honey with minimal disruption to the nest. It is particularly beneficial not to disrupt the brood. Honey hives are also suitable for propagation by division. Honey hives work well even if you don't intend to extract honey, as they have better insulation on the top and a storage area that can be used when times are good but easily sealed off when conditions are poor. For this reason, they also make good pollination units.

The Honey OATH design was developed for *T. carbonaria* and *T. hockingsi* in the subtropics. See the discussion of mini-hive designs in the following section for honey production from smaller species.

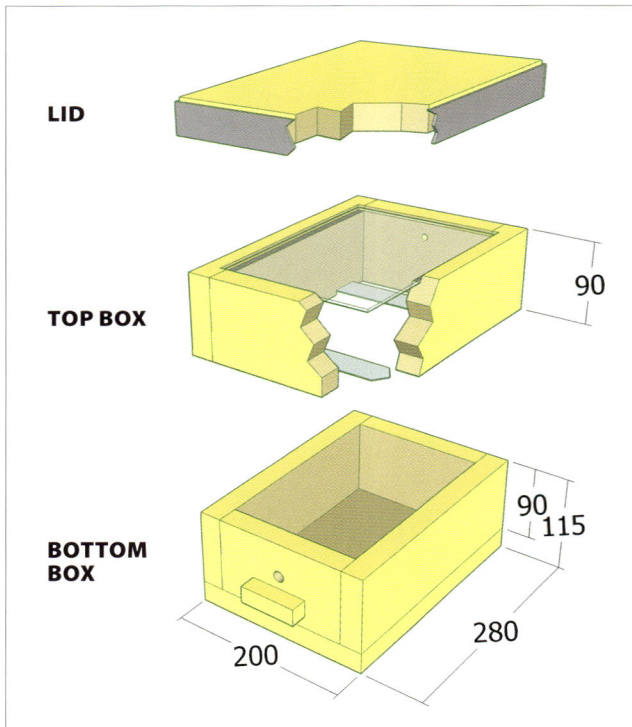

FIGURE 9-16
Three-dimensional diagram of a two-section OATH hive showing observation window in mid box.
DRAWINGS **DANIEL KLAER**

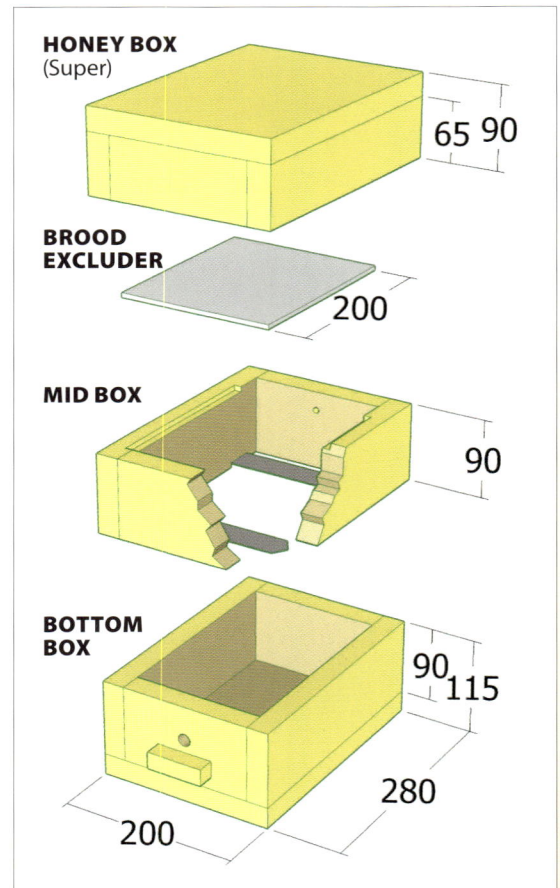

FIGURE 9-17
Details of a three-section Honey OATH hive showing the top honey section and the brood excluder.
DRAWINGS **DANIEL KLAER**

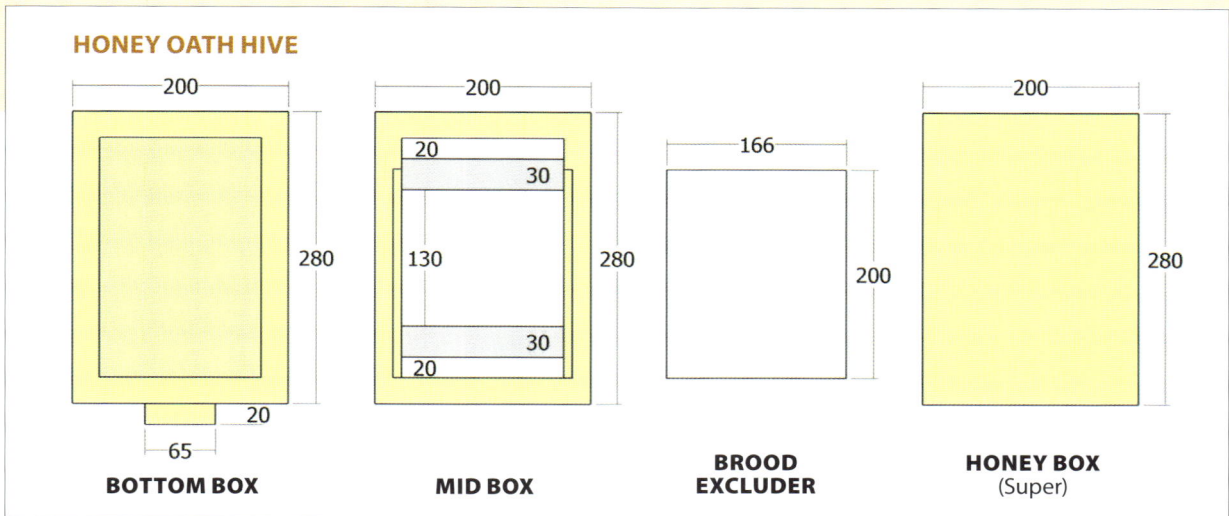

HONEY OATH HIVE

200	200	166	200
BOTTOM BOX	**MID BOX**	**BROOD EXCLUDER**	**HONEY BOX** (Super)

FIGURE 9-18 Plan of a three-section Honey OATH hive, illustrating the split bars and brood excluder. DRAWINGS **DANIEL KLAER**

This honey hive design consists of a standard OATH hive without a top board, with a honey section 45–70 mm deep on top. The rebated brood excluder separates the honey section from the rest of the hive and prevents the brood from entering this section of the hive. A slot at one end of the excluder allows the worker bees to enter the honey section to store their honey. The brood excluder can be made of thin wood or plywood, but I prefer a 5 mm-thick plastic called HDPE (high-density polyethylene), which is readily available in the form of breadboards. It needs to be secured into place (Figure 9-19).

FIGURE 9-19 A brood excluder nailed securely into the top of the middle section of a three-section Honey OATH, with a slot for entry of workers.

The variability in depth of the honey section from 45 mm to 70 mm changes the volume from 1.6 litres to 2.4 litres. Any depth within this range has been shown to work well and be acceptable to the bees for honey storage. Making the honey section deeper increases the chances that the bees will seal off the space. Making it too small fails to exploit the full capacity of a hive to produce honey. I prefer a honey section that is 65 mm deep (volume 2.2 litres, total hive volume 8.4 litres).

The slot allows the bees to easily extend their peripheral space into the honey section. The peripheral space is both a ventilation duct and a transport route that the bees construct around the outside of the nest close to the hive wall (Figure 4-16). The use of a slot allows the bees to extend their peripheral space and therefore encourages them into the top section where we want them to store lots of their delightful honey (Figure 9-20). An alternative design uses a hole, or series of holes, drilled through the brood excluder, which seems to also work well. But, if you do this, make sure to keep the holes away from the middle of the excluder or the brood may enter.

The connection between the honey section and the brood sections is deliberately placed at the back of the hive. It works better there than at the front because honey is naturally stored at the back of the hive (Figure 4-17). The connection at the back encourages the storage of honey and not pollen in the honey section. (Read more about this in "Food storage pots" in Chapter 4.)

Constructing hive boxes for stingless bees • 129

A honey hive can be divided to form two hives just like a normal OATH. I do not recommend that you remove honey at the same time as dividing because this places excessive stress on the hive. Instead, wait at least a week after division and then collect the honey. Or you can extract honey and divide the hive at least one week later. If the original hive was full, you should be able to extract honey from the original honey section. The hive resulting from the division (which has the full bottom section) will not refill its honey section with honey for some time — about one year under favourable conditions. (Read more about how much honey a hive can produce in "Productivity of hives" in Chapter 12.)

FIGURE 9-21 An insert honey section that fits below the normal honey section of a hive and gives extra capacity to store honey.

9.4.4 MINI OATH HIVES

Mini hives are tailored for stingless bee species with smaller nest volumes (Table 7-2). Mini OATH hives have the external footprint dimensions of 200 mm X 200 mm (Figure 9-22). They can be made as a normal two-section hive or with a honey section on top (Figure 9-23, Figure 9-24).

I recommend mini OATH section heights of 60 mm. Assuming a wall thickness of 25 mm, and depending on the exact height, the internal volume of a two-section mini OATH and a three-section mini Honey OATH is 3 litres or 4 litres respectively, which is slightly less than half the internal volume of a normal hive. Regular hole diameters for the entrance and vent are 10 mm and 6 mm respectively.

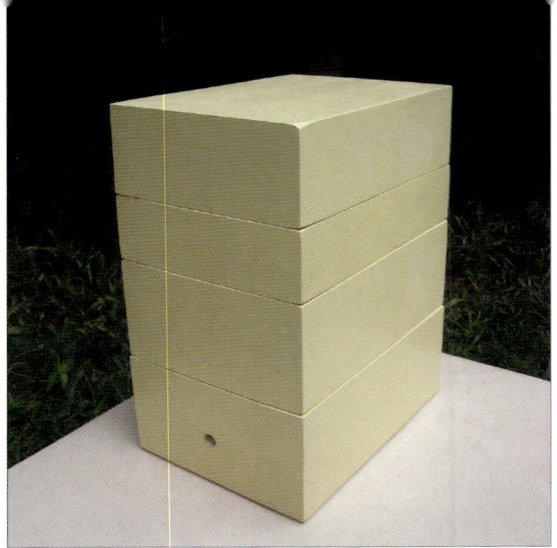

FIGURE 9-20
A full honey section removed from the section below, containing at least 1 kg of harvestable honey.

IMAGE **KATINA HEARD**

A question that remains to be answered is the optimum ratio of the volume of the brood section to the honey section. In the standard honey hive (Figure 9-18), the brood box is 6.2 litres and the honey section 2.2 litres. The large brood box provides the bees with ample reserves even after the honey has been extracted from the honey section. So, even if the bees face a tough time ahead, their survival is ensured. But it may be possible to extract a greater proportion of the hive reserves. This will allow for more profitable harvesting of honey, but could decrease survival of the colony through starvation.

It is possible to insert a second honey section between the brood box and the top honey section (Figure 9-21). I call these "insert honey sections". They give more storage space, meaning that you do not have to extract the honey so often. Perhaps, instead of every six months, you can extract every 12 months. An added advantage of the insert honey section is that the colony may be able to perceive that there is space in the hive ready to be filled. This motivates them to forage more, resulting in more yummy honey for you!

FIGURE 9-23 Mini OATH with observation window *(LEFT)* • Mini Honey OATH *(RIGHT)*.

MINI OATH HIVE

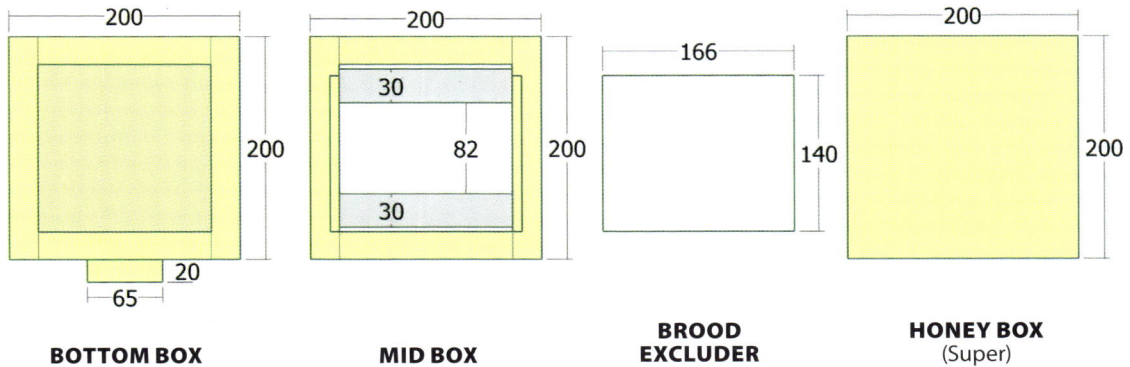

BOTTOM BOX — 200 × 200, 65, 20

MID BOX — 200, 30, 82, 30

BROOD EXCLUDER — 166 × 140

HONEY BOX (Super) — 200 × 200

FIGURE 9-22 Plan of a three-section mini Honey OATH, illustrating the split bars and brood excluder.
DRAWINGS **DANIEL KLAER**

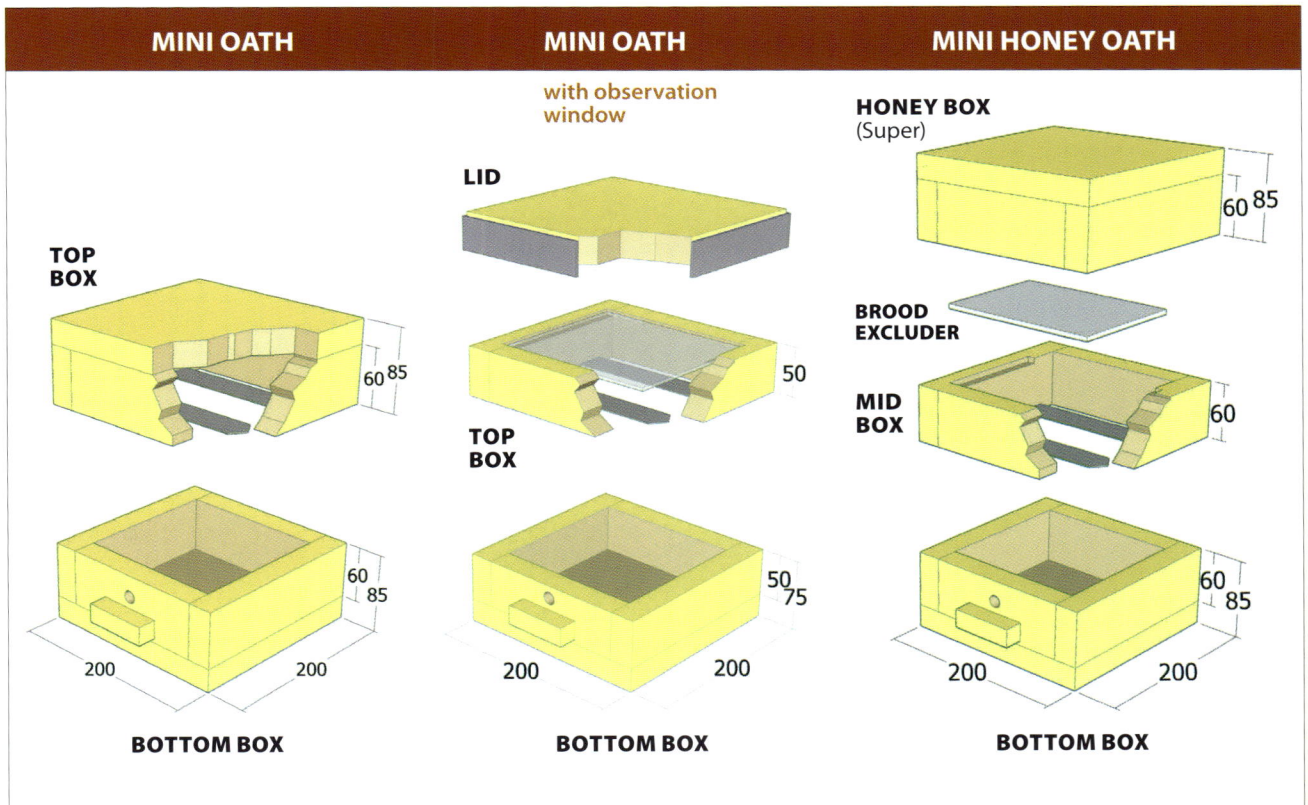

MINI OATH

TOP BOX — 60, 85

BOTTOM BOX — 60, 85, 200, 200

MINI OATH

with observation window

LID

TOP BOX — 50, 50, 75, 200, 200

BOTTOM BOX

MINI HONEY OATH

HONEY BOX (Super) — 60, 85

BROOD EXCLUDER

MID BOX — 60

BOTTOM BOX — 60, 85, 200, 200

FIGURE 9-24 Details of mini OATH hives *(FROM LEFT TO RIGHT:* OATH, OATH with observation window, Honey OATH), showing dimensions, split bars, and other features.
DRAWINGS **DANIEL KLAER**

10 Establishing a stingless bee colony in a hive box

Removing a nest from a log for transfer to a hive.
IMAGE **DAN COUGHLAN**

Stingless bees can be **kept in their original log,** but you need to protect it.

If the log deteriorates, it is **best to transfer the colony** into a hive box; once it is transferred, you can move it, propagate it and extract honey.

You need to **follow some simple guidelines** to help the transferred colony to thrive.

If a colony is **difficult to access** and so can't be transferred, you can try to bud a colony from it instead.

Now that you have made your box, you need to get yourself a colony of bees! You can establish a colony in a hive by means such as transfer, budding and trap hives, all of which I discuss in this chapter.

First, we will talk about how to find a colony. Then I would like you, dear reader, to consider whether it is really in the interests of the colony to move it, or if you should just leave it be. I go on to present some tips on how to keep a colony safely in its original log, which I encourage if the log is in good condition. If the colony would clearly benefit from being rescued, you will also learn a few ways of doing this. This includes learning the classic transfer procedure, in which an original nest site is opened and its contents moved to a new home in one swift operation.

I complete this chapter by discussing two other methods of founding colonies: 1) budding and 2) trap hives. Budding is a way of duplicating a parent colony by attaching a hive box to the parent colony entrance and waiting for a second colony to form in the box. Budding is a moderately successful method under favourable circumstances. The second method, laying out trap hive boxes to attract a naturally founding colony, sounds logical, but has a disappointing success rate. But I'm happy to share with you the one trick that saves this technique from being a total failure.

"Mini-colonies" is a third method that you may hear about. This refers to a strategy of transferring a quantity of brood and food from a thriving colony to a small-volume hive. However, it has not worked well for Australian species, so I do not discuss this.

10.1 How do you find a wild stingless bee colony?

The most common natural nest site for stingless bees is a hollow log. In artificial environments, colonies can often be found in wall cavities, under concrete slabs, in compost bins, and in underground service boxes such as those used for water meters.

Finding a wild colony is not easy. Learning how to find them yourself may take years of practice and patience. Look for the foragers early in the morning or in the late afternoon when the sun is not in your eyes. Alternatively, look on cloudy days when foragers can be more easily seen against the dull grey background.

Bees are not fussy about tree species and will nest in any type of tree with a suitable hollow. They may nest high up, and be difficult to see, so focus on lower parts of trees where you are more likely to see them. Look for the glistening of resin around the nest entrance, and for knots that may surround entrances into a tree hollow. Also look for swarming behaviour, which is more obvious.

Another search strategy is to set out feeders of sugar water to attract bees and determine their presence that way. You can also learn bees' natural enemies and look for signs of those. Bembix wasps and hive syrphid flies may be seen around nests and are larger and more conspicuous than the bees themselves. (Read more about these pests in Chapter 14 "Natural enemies of stingless bees".)

One final strategy to find the location of nests in natural or artificial locations is to ask around or advertise in the media. Solicit the help of foresters, arborists, clearing contractors and millers. For example, you can ask them to cut and keep the relevant section of any log that contains a nest. Collect it later and offer a reward. (Read more about how to move a log in "How to transfer a stingless bee colony to a hive box" later in this chapter.)

Remember that it is illegal and unethical to remove anything from a conservation area. Even on private land, I certainly do not recommend that a tree be chopped down or damaged to remove a bee colony.

10.2 Keeping a stingless bee colony in a log

If bees are living securely in a solid log, then I recommend that they be left there. The log provides a great home for the bees and an attractive feature for a garden.

But you need to protect their home. Keep it above the ground so termites are deterred from entering. Keep it dry to slow down the rotting process. Cover the open ends of the log to help the bees defend and insulate their nest. You can do this by nailing or screwing a thick timber or plywood plate to each end of the log. Timber is a good insulator, but the top plate will need further protection from sun and rain so cover it with a metal plate (Figure 10-1). Any large gaps in the log that may allow pests to enter are best sealed with a wood filler.

You can lie the log horizontally or stand it vertically in its new location. You can also change

FIGURE 10-1 Stingless bee colonies suitably housed in hollow logs. BOTTOM IMAGE **GLENBO CRAIG**

its orientation from horizontal to vertical if necessary for transport or mounting, but do not do it often as it disrupts the rearing of the brood. Ants may cohabit with bees in the log. It may appear that they are living in the same space as the bees, but bees normally seal off their space so that ants cannot enter. So do not worry too much about ants; they may even help by deterring termites, which will eventually destroy the log.

10.3 How to transfer a stingless bee colony to a hive box

Colonies of stingless bees can be moved from their natural sites by opening their nest and moving the contents, including the brood, food and bees, into an artificial hive box in one operation. This is a moderately quick and easy technique with a high rate of success.

10.3.1 WHY TRANSFER A STINGLESS BEE COLONY?

When a log containing a nest has been damaged or is rotten, the transfer of the colony into a hive can save them (Figure 10-2). It is also much easier to transfer a colony from a rotten or damaged log compared to a solid log. You should only need hand tools such as an axe and wedges. My rule of thumb is that if you need to use a chainsaw or power tools, you should not be moving the colony, as the log is in good enough condition to provide a good home.

Moving a colony also makes sense if it is under threat because it is not welcome in its current location, such as in a water meter box (Figure 10-3). In this case, you can rescue the bees by transferring the colony into a hive.

FIGURE 10-2 This colony was in the log on the right that fell and broke open. The colony was in extreme danger but was rescued by successfully transferring it into a hive box. IMAGE **JEFF WILLMER**

FIGURE 10-3
These colonies of *Tetragonula hockingsi* in subterranean utility boxes had to be moved, so transfer to a box was a means of rescuing them.

A further advantage of transferring a colony to a hive is that you open up new opportunities for propagating and utilising the bees. A beekeeper can divide the colony to create more colonies. The hive can be opened to extract honey. And hives are more easily sealed and transported to another location, such as a farm for pollination or to new owners for pets.

10.3.2 HOW TO TRANSFER A STINGLESS BEE COLONY

If you are in an emergency rescue situation, then the log may need to be moved and stabilised. Wrap it in multiple layers of cling wrap to keep the bees in and pests out. Try to move it at night or in a cold part of the day when the bees are all inside. Try also to keep it in its original orientation, but do not worry if you need to change it. Place it in your vehicle on carpet or cushions to absorb shocks.

When you reach your destination, move the log as carefully as possible to a safe location. The best location will be close to your home, in the shade, and sheltered from rain. Raise the log off the ground to help protect it from ants and termites. Once it is in position, open the entrance by puncturing a hole in the cling wrap. Wait a few days before transferring the colony into a hive box, so that foragers have learnt their new position and will orientate to the box's entrance more easily.

◀

FIGURE 10-4
The log has been opened by splitting it down two sides and separating the two halves. The brood is being removed ready to place in the box.

FIGURE 10-5 A tray is useful to hold nest material during the transfer. Any remaining honey pots or spilled honey can be enjoyed later.
IMAGE **CLAUDIA RASCHE**

FIGURE 10-6 A brood chamber removed from a log in excellent condition, still surrounded by its involucrum.

FIGURE 10-7 Nest contents in the hive box, including the brood in the middle and nest material and food pots around the outside.

Have the following gear ready:

- tools to open the log: axe, hatchet, hammer, wedges, wrecking bar, saw
- tools to remove the nest contents: hand spade, large sharp knife, small knife
- other equipment: hive box, strap to secure hive parts together, tape to seal box, lump of propolis to modify the hive entrance, tray to temporarily hold nest contents, colander, container with lid to hold any materials to keep for later, and bucket of water and rags to wash sticky honey off box, hands and tools.

Prepare the hive box:

1 Drill a hole about 7 mm in diameter in the bottom section of the box to allow spilled honey to drain. I prefer to place the hole in the back corner.

2 Insert a wire mesh false floor in the bottom of the box. Fold the edges of the wire so that it sits above the floor of the box (Figure 10-4).

3 Make a propolis ring (obtained earlier from another hive or salvaged from the nest being transferred) and press it around the outside of the hive entrance to reduce the diameter from the normal 13 mm to about 5 mm. This provides a strong visual cue for the bees to find their new home. Do the same on the inside of the hive entrance to form an internal defensive nest entrance tube (Figure 14-21).

Use appropriate tools (axe etc.) to open the log or other cavity to get wide access to the nest (Figure 10-4). Move the nest contents into the box (Figure 10-5). The brood chamber and the workers are the most important components to transfer. Remove the brood carefully to reduce damage and keep it as intact as possible. Protect the involucrum surrounding the brood and keep it as intact as possible; Figure 10-6 shows an intact involucrum. Put the brood aside in a container in a safe position. Try to keep the brood in its original orientation.

Select some nest material that includes propolis structures and intact food pots. (Include only soft, fresh propolis that workers can mould into new structures; older, harder propolis is not much use to them.) Place this nest material in the bottom of the hive box around the four corners to create a cradle for the brood chamber. Place the brood chamber on this cradle (Figure 10-7). If the brood falls apart, then gently arrange it in the box, ideally with the advancing front at the top.

Stored food is valuable to sustain the colony while it re-establishes. It is tempting to transfer large quantities, but about 200 mL is plenty to tide the colony over. You need to be particularly careful of broken food pots. Spilled honey will drown bees and broken pollen pots can become infested with mould. Broken pollen and honey pots will also attract natural enemies. Place the food stores around the brood, not on top, or the heavy food will crush the delicate brood.

You may have material left over after a transfer. You can keep this in the refrigerator and add it later to this hive or another. Another option is to leave it near the hive and allow foragers to move it into their new hive at their own pace. I find that they are mainly interested in the propolis, occasionally interested in honey, but not likely to scavenge the pollen at all.

For the *Tetragonula* species, you do not need to worry about the queen, which is almost always present in the brood chamber. But, in the case of *Austroplebeia*, which does not encase the brood with an involucrum, the queen may flee to a remote part of the nest cavity. You need to search for her and reunite her with the brood to maximise the chances of a successful transfer. An **aspirator** (also known as a "pooter") is a useful tool to catch the queen but, if you are gentle, you can safely grab her with your fingers. Peter Davenport was

transferring an *A. australis* and noted that the queen was mired in honey. He gently popped her into his mouth for a minute to clean her up then placed her in the hive box with the brood, where she carried on as if nothing had happened!

Once you have transferred all the necessary material to the hive box, put the top section on. Strap the hive sections together (Figure 10-8). Seal the gaps with tape. Place the box as close as possible to the original position of the log. If you do this, any stray bees will quickly find their way home to the hive box (Figure 10-9). A large population of bees is crucial for the success of the transfer operation.

Tilt the box so that the drainage hole is the lowest point to drain off any spilled honey. Place a collecting bowl under the hole to collect and remove the spilled honey, most of which will be drained within a few hours.

FIGURE 10-8 The hive has been closed and sealed. Propolis has been placed around the entrance.

If the nest being transferred is small and fits entirely into the bottom section of a multi-section hive, then you can place a board, or clear plastic sheet, over this section so that your bees establish in this small space first. Later, the upper section of the hive can be added. Do not open your hive for at least two weeks after transfer as this may compromise the seal between the hive sections and allow the entry of natural enemies. A hive with an observation window will allow you to see what is going on inside.

After a few months, the bees will have established in the box, covered the brood, built the internal entrance tube, cleaned up any spilled honey and repaired broken food pots. They will have built numerous connectives to join the nest parts together and to join these parts to the wall of the hive so all is secure inside (Figure 10-11).

You have two options with regard to the original nest site. You can destroy it and remove it so that stray bees are not tempted to return and are more likely to find and move into their new home.

FIGURE 10-9 Following the transfer of a colony into a box, stray bees orientate to their new home. IMAGE **GLENBO CRAIG**

When a log colony is very large and strong, it may be possible to create two colonies from the original one. It is usually preferable to transfer it into one box so that the maximum number of workers is available to establish the hive. However, if the nest does not fit into a single box, the brood chamber can easily be divided into two, and there are plenty of worker bees present, you can try to transfer the colony and nest into two hive boxes (Figure 10-10).

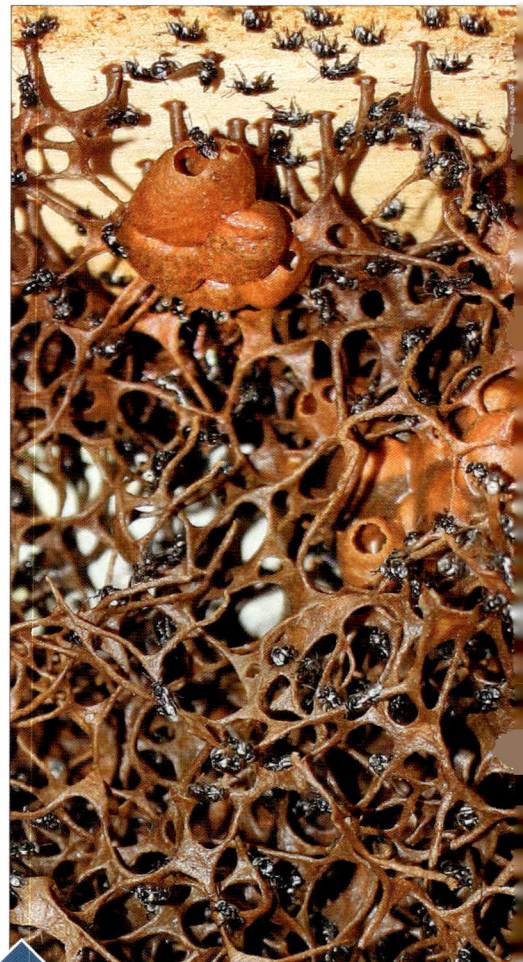

FIGURE 10-10 Two hives from one large nest of *Tetragonula hockingsi* found in the water meter box seen underneath the hive boxes. The image shows two bottom box sections almost full of brood, food pots and propolis structures prior to the top sections being added. Both of these colonies successfully established. IMAGE **CLAUDIA RASCHE**

FIGURE 10-11 After a few weeks or months, the colony has established in the hive box. They have covered the brood, built their internal entrance tube, sealed gaps, repaired broken food pots, and connected the nest parts to each other and to the walls of the hive. IMAGE **GLENBO CRAIG**

Do this after striking each bit to dislodge the bees. The advantage of this strategy is that you maximise the number of bees in the one location. The colony needs workers to repair all the damage caused by the transfer. The more workers you can entice into the hive, the bigger the workforce to quickly make the necessary repairs. John Klumpp recommends that lost workers be collected using an aspirator and placed in the hive.

Another very successful and useful strategy is to put the log back together to allow it to regenerate. I call this the "Tony Goodrich technique", after my great friend and long-time keeper of stingless bees who showed it to me. Ideally, leave some brood in the log, although this technique can be successful even when no brood is left. Put the pieces of log back together and wrap them in cling wrap. Use the cling wrap plentifully to get multiple layers around the log to protect it (Figure 10-8). Place the log back in its original location close to the hive box so that the two new homes attract any stray bees. In the majority of cases, both colonies will survive.

The Tony Goodrich discovery is significant, because we effectively get two colonies where we had one and, what's more, it can be repeated each year. Tony has logs from which he transfers a colony annually. They survive the process and live to spawn another colony next year. A downside of this technique is that the workers are divided between the two new colonies. It is probably not wise to attempt this if the worker population is not strong or if other factors (e.g. weather) are not ideal. However, if the colony is so strong that it cannot fit into one box, then this is an excellent choice.

Place the hive (and the log if you are using the Goodrich technique) as close as possible to the log's original position so that the bees find the new home quickly.

If you are creating two new homes, a new queen is needed for the new location that lacks a mated queen. We expect the original queen will now be in the new hive. After all, we transferred the entire brood chamber and that is where she spends most, if not all, of her time. But somehow both colonies usually manage to pull through. (Read more about this in "Queen replacement" in Chapter 3.)

Hives that have been recently transferred or split are temporarily more susceptible to invasion by natural enemies. This is because of the formation of entry points for those enemies, the damage to food stores where enemies often start their development, and the loss of bees available for defence. Gaps between the hive sections may allow some enemies to enter (particularly hive syrphid fly). The entrance may be unprotected because of the lack of the defensive internal entrance tube. Small hive beetle and hive phorid flies are the enemies most likely to enter through an unprotected entrance. To prevent them from damaging your colony, it is crucial that you read and heed the lessons in "Protecting your hives from natural enemies" in Chapter 14.

10.3.3 TRANSFER OF A COLONY BACK INTO A LOG

Natural log nests are a very attractive garden feature and so are in high demand. But they are becoming increasingly difficult to obtain. Accordingly, more people are interested in learning how to stock a natural log with a colony from a box.

It is feasible to move part of the contents of a hive into a log and still retain the hive. First, find a suitable hollow log. The volume of the internal space should be close to the natural nest volume of the bee species (Table 7-2). The walls of the log should be at least 25 mm thick. If the log lacks a natural entrance hole, then drill a 13 mm hole.

Now, follow the instructions for preparing your natural bee home outlined in "Keeping a stingless bee colony in a log", but leave one end open for access. The transfer will be easier if the log is lying down. Open a full hive. Transfer half of the brood into the log. The brood can be taken from either the top or the bottom section of the hive, but it is easier to securely place the bottom half of the brood with its surrounding involucrum underneath. Also transfer some nest material, including a quantity of food provisions. Read the above section "How to transfer a stingless bee colony", and apply the guidelines in that section to the reverse process. Close the end of the log with the end plate and fasten it.

10.4 Budding (duplication, eduction or soft splitting)

First discovered by Tom Carter and further developed by John Klumpp, the budding technique is very useful for founding a new "bud colony" from an existing parent colony in a location where you cannot gain access to remove it. The location may be a tree hollow or a house wall, such as the one in the following example. The parent colony is not removed from its original location but continues to live there. Therefore, it can be repeated yearly if the parent hive is strong. This technique works for species of *Tetragonula* and *Austroplebeia*.

Mount an empty hive box close to the parent colony (Figure 10-12). The hive box should not be too big; the bottom section of an OATH hive is ideal. The box needs to have an observation window in the top so that you can monitor the proceedings inside. (In the images below, an acetate sheet has been taped to the top of a box to form the observation window.) The window needs to be covered with a removable plate to exclude light. Connect the entrance of the parent colony to the back of the hive box with a tube. Black poly irrigation pipe and fitting work well and are easy to find. Insulate the connecting tube and make it opaque to light (in the example, the inner tube is covered by a length of hose pipe secured in place with black gaffer tape). The bud box is now in place and the foraging bees are forced to move through it to access their nest. Provide shade if the bud box is in full sun.

The bud box may be accepted as part of the nest and food pots built there (Figure 10-13). Eventually, the connected hive box may acquire a queen, likely a virgin queen from the colony to which it is attached. The virgin queen then mates and begins to lay eggs in the brood cells built by

the workers. This rebel bud colony is now at risk. If the parent colony recognises it, they will destroy it. Hence weekly inspections are needed and, when a brood is seen forming, the bud hive must be disconnected from the parent colony. John Klumpp recommends a stage of partial disconnection before final total disconnection. Do this by cutting a hole or inserting a T-piece into the connection between the parent nest and bud box. Partial disconnection seems to allow the parent colony to continue to contribute workers to the bud colony but reduces the chances of the parent colony killing the bud colony queen.

In addition to budding a colony from an inaccessible location, a few beekeepers use this method to propagate hives as an alternative to the classic OATH division (Chapter 11). This certainly has the benefit of avoiding ruptured honey and pollen pots that attract the attention of natural enemies. But it also takes months instead of minutes, needs weekly inspections and, on many occasions, is unsuccessful because the bud colony does not form at all.

Tony Goodrich simplifies the budding process. He sets up the bud box as above. At four months, he inserts a T-piece into the connection between the parent nest and bud box. After another four months, he disconnects the two nests. After another four months, he takes the bud box away. This calendar method is useful where it is not possible to do weekly inspections.

W1　　　W4　　　W8

Budding

FIGURE 10-12

The set-up used to bud a stingless bee colony (the original colony) from an inaccessible location into a hive box (the bud box).

A The site of the original colony.
B Plate glued on the entrance of the parent nest.
C Connecting tube being inserted into parent nest entrance.
D Hive being connected to parent nest.
E Parent nest connected to bud box (without lid).
F Bud box with lid and roof on.

FIGURE 10-13 The development of a bud hive over 45 weeks. **W1** (Week 1) – **W12**: the bees fill the bud hive with propolis structures and food stores. **W15**: the bees have built the first brood in the bud hive. **W16**: the bud hive is disconnected from parent hive. **W45A**: the bees have completely filled the bud hive with stores and built a large brood. **W45B**: top sections are added to hive to complete the process.

W 12 **W 15** **W 16** **W 45A** **W 45B**

10.5 Trap hives

Unfortunately, there is only a very low probability of capturing a colony in a trap hive

There have been many attempts by beekeepers and a few scientific studies designed to attract and capture a stingless bee colony into artificial boxes, called trap hives or trap nests. Keepers of honey bees sometimes use this technique, usually baiting the hive box with a pheromone lure. For example, the control of the Asian honey bee in north Queensland uses trap hives that are then destroyed. Therefore, it seems logical to leave empty trap hives out to attract a founding colony of stingless bees.

Unfortunately, there is only a very low probability of capturing a colony this way. Attempts in Asia and South America have had limited success. For example, one thorough study in Sumatra showed that of the 362 trap hives left out for more than four years, only 6% were colonised and only 1 species out of the 24 present in the area (*Tetragonula minangkabau*) occupied the trap hives. In South America, empty soft drink bottles are sometimes colonised but, as in Asia, at a low rate and by few species.

In Australia, there are a few reports of a trap hive being occupied by *Austroplebeia* species bees. Peter Davenport has directly observed one successful occupation. It is even rarer for

FIGURE 10-14 A fighting swarm at the entrance of a *Tetragonula* hive.

Tetragonula species. It seems that empty boxes are not recognised or accepted by scout bees searching for a location to found a new nest. Perhaps the search cues that bees instinctively use to find and identify a suitable location are not satisfied by the artificial appearance, smell, texture and/or conditions of a hive box.

However, there are circumstances in which you have a chance of capturing a swarm of bees in a box. Stingless bees can sometimes be seen flying in an aggregation near the entrance of the existing hive (Figure 10-14). This may be an attacking swarm attempting to take over an existing colony (read more about fighting swarms in Chapter 13). You can simultaneously divert the attackers away from the defending hive and capture the invaders by following a simple process.

Remove the defending hive to another site, following the guidelines outlined in "Moving your hive" in Chapter 8. Place the hive that you want occupied in the position where the attackers are focussing. Attacking swarms do not normally raid hives unless they contain an existing colony. But, if the invading swarm has been "blooded" by their previous attempts, they may proceed with the invasion. If you happen to have a weak or recently dead hive, then you may be able to capture the swarm into this one and end up with a strong hive. The weak or dead hive has a lot of useful resources in the box, especially its entrance tube, propolis structures and stored pollen and honey. So, instead of cleaning this box for reuse, put it in the place of the invaded hive. Eventually the invaders may establish their new colony in there by installing their queen and starting to rear brood.

If you do not have a weak or dead hive, try using a recycled hive. A recycled hive is one that has been used before but the bees have died and it has been cleaned out for reuse. (Cleaning is often required because old hives may contain natural enemies that have moved into an undefended hive.) Ideally, leave a layer of propolis on the inside surfaces. Also leave the entrance tube that the bees construct behind the entrance hole.

If you have only a new trap hive box, prepare it by melting and pouring propolis and wax from another hive around the inside of the new box. Make an entrance ring and squash it into place.

Transfer
FEATURE

Transfer of a colony of *Tetragonula carbonaria* from a log to a box.

11 Splitting stingless bee hives

Thousands of hives have been generated annually in Australia by splitting (dividing) existing hives.

The **standard splitting technique** involves separating two sections of a full hive and coupling each half with a new empty hive.

The **success rate** of splitting hives is very high if your box conforms to a few basic design rules, is built well, and you follow the guidelines.

Most Australian stingless bee species are **able to be split** using the standard OATH design and technique.

Colonies of Australian stingless bee species **re-queen** themselves if a hive division renders them queenless.

IMAGE
DAMIEN ANDREWS

11.1 Introduction and history

The development of artificial division by splitting of stingless bee hives has been the impetus behind the explosion of interest in these bees in Australia since the 1990s. Before this, bee colonies could be kept in logs, even transferred to boxes, but they could not be split. Propagation by splitting existing hives has led to exponential growth in the number of hives and increased interest in keeping them.

Geoff Monteith and Sybil Curtis, from the Queensland Museum, discovered this technique. In the 1970s, their existing small box hive had been enlarged by putting an extra box on top and the colony had expanded to fill the double volume. When Geoff's father, Jock, suggested they could split it in two by cutting through the join with a long knife and adding new half boxes to the two sections, they reluctantly did so, fearing that one half would die without the queen. Jock took one of the new hives home and, to their surprise, both prospered. This basic method has not changed much; hive division still requires separating two sections of a complete hive and then coupling each half with an empty half to create two complete hives from one. The original two halves provide a nucleus for each of the new colonies (Figure 11-1). This horizontal split technique is a key feature of the Original Australian Trigona Hive (OATH). (Read more about the OATH in Chapter 9 "Constructing hive boxes for stingless bees".)

Years after Geoff and Sybil discovered this method, I was visiting a remote indigenous community of keepers of stingless bees in Mexico, and was astonished to find that they divide their clay pot hives in a similar way. But they independently developed the method centuries (or millennia) before us. (Read more about Mexican beekeepers in the feature "Indigenous peoples and stingless bees".)

The success rate of splitting hives is very high if your box conforms to a few basic design rules, is built well, and you follow the guidelines. I have recorded the outcome of splitting over 1000 hives and, in 94% of cases, both halves have survived. In 6% of cases, one half died. Both halves died in only 0.3% of cases.

FIGURE 11-1 This OATH hive has been opened ready to be divided into two halves. The top section, on the left, has been inverted. The bottom section, on the right, is in its normal orientation. Each half is fully occupied with brood, stored food and adult bees, and so makes an excellent nucleus for two new hives. Note the two split bars in the top section, which aid separation and prevent the contents from slumping into the bottom section.

11.2 Why divide your hive?

There is a huge demand for stingless bees as pets, pollinators, honey producers, and educational resources. Propagating hives provides a way to meet that demand. Even if you do not aspire to breed lots of hives, all beekeepers should keep at least two hives; then, if one colony dies, you have not completely lost your bees, but can propagate from the remaining one. Once you have two hives, you can extract honey or keep dividing to provide hives for all your friends who will want one when they see yours.

On the other hand, you should not feel an obligation to divide your colony of bees. The colony will continue to live for many years undivided. And it will probably attempt to reproduce naturally by founding a new colony if a nearby suitable location is available. If this happens, do not fear, you will not lose your colony, as the parent colony will remain in its original location. If it is your intent to allow your colony to naturally found new nests around the hive, then it is a good strategy not to split the hive. This may be desirable in disturbed areas where bees have declined. But, if the hive is in an urban area, then the new colonies may be founded in undesirable locations such as water meter boxes and compost bins, where they could be a nuisance. Dividing the colony artificially is the most certain way of enhancing their populations.

Not splitting on a regular basis may also make it hard to split your hive later on. Peter Clarke has split hives that have not been divided for more than six years, and found that the honey and pollen pots had assumed a hard consistency, making splitting very difficult.

11.3 When to divide a hive

Determine that the hive is ready for division by estimating its weight, time since last division, and forager activity. Determine the gross weight with bathroom scales on a firm flat surface and subtract the weight of the box to estimate the net weight (Figure 11-2). *Tetragonula carbonaria* colonies on the subtropical east coast of Australia, for example, can be split when they reach around 3 kg net weight, which is typically achieved around 12 months after the last split. By this time, the colony will normally have expanded to fully occupy the hive volume. Forager activity should also be strong. In general, for *T. carbonaria* and *T. hockingsi*, you should be able to observe between 30 and 60 bees returning to the hive per minute when environmental conditions are favourable. (Read more about how to estimate your hive's strength in Chapter 13, Section 13.4).

FIGURE 11-2 Bathroom scales are accurate enough to weigh a hive of stingless bees
IMAGE **MIKAYLA LAMBERT**

Ideally, split your hive when weather conditions are moderate. Do not do it when it is raining unless both hives will be kept under cover. Rainwater can enter the gap between the sections of the box if the bees have not had time to completely seal them. Water in the hive can be a problem.

It stimulates the growth of fungus, which can destroy the colony. Also do not divide the hive if you are expecting cold or extremely hot weather. Bees need time to rebuild their insulating structures after a division. I divide hives of *T. carbonaria* all the year round in coastal south-east Queensland, including in winter when the temperature fluctuates between a daily minimum of 10°C and a maximum of 22°C. But it is probably not a good idea to divide hives in winter if you are situated farther south or inland where temperatures are lower. Hot weather is generally not a problem, but avoid splitting in extremely hot weather, above 40°C.

Hives can be split at any time of the day. Some beekeepers prefer night splitting because bees are less likely to fly then, but be careful as bees may walk out of the open box in large numbers and risk being crushed or lost.

Please note that, the further south you go, the longer it will take for your hive to recover from splitting and be ready to split again. South or west of Sydney, your hive may never reach sufficient weight to be able to split it.

11.4 How to divide a hive

First, you may wish to provide some protection for yourself against the bees. (Read more about this in "Protecting yourself against the defences of stingless bees" in Chapter 13.) After dividing your hive, you will also need to protect it from natural enemies. (Read more about this in Chapter 14, section 14.11).

Have ready the following items: a compatible empty hive, hive tool, small knife, and tape or straps to fasten hives after the split. Figure 11-3 shows the process for *T. carbonaria* bees in a three-section Honey OATH, but the principle is the same for a two-section OATH.

Separate the top and bottom sections of the box with a sharp tool such as an American hive tool or heavy-duty paint scraper (Figure 11-3). Some effort may be required because the bees stick the sections together with propolis. Some beekeepers simply pull apart the two sections, while others make a horizontal cut to facilitate the separation. A complete cut between the two halves would damage the brood, so only cut in the corners

FIGURE 11-3 **A** The hive of *Tetragonula carbonaria* on the right is full and ready to be divided. An empty box section on the left will be used for the division.
B The full hive is opened using a hive tool.
C The two sections of the hive are separated.
D The full top is coupled with the empty bottom.
E A rag is used to clean spilled honey from the flat joining surfaces.
F The full bottom is coupled with the empty two top sections.
IMAGES **MIKAYLA LAMBERT**

where the food pots lie. If split bars are fitted to the boxes, then cutting is not usually necessary. After inspecting the brood, it may be necessary to make small strategic cuts (Read more about this in "Various brood separation scenarios", later in this chapter.)

Use the hive tool to scrape any obstructions from the joining surface of the boxes. Use a rag to clean any spilled honey from the flat joining surfaces of the box sections to reduce entry of natural enemies. Couple the full top with an empty bottom section. Couple the full bottom with an empty top section (Figure 11-3). Check that the boxes come together closely, leaving minimal gaps. If gaps persist, then run tape around the join between the boxes.

Secure the hive parts together. You can use adhesive tape either horizontally or vertically around the gap between the sections. Tape is readily available and works well but leaves unsightly glue marks. A simple and effective fastener is plastic strapping and buckles (Figure 11-4). These are cheap and strong. They will deteriorate in under one year in the sun but, by then, the bees will have glued the hive sections together. They are not easy to buy in small quantities but most beekeepers have plenty and will happily part with a few. More durable options include cable ties, tie-down straps, or a Zabel's metal mini-emlock fastener.

FIGURE 11-4

Fastening the plastic strapping to hold the hive sections together.

INVERTING A HIVE

Note that we maintain the hives in their original orientation during the division. The common wisdom is that you should not invert comb. This argument is entirely plausible. Eggs and young larvae float on semi-liquid provisions (Figure 4-32). If you turn the hive upside down, the provisions could run to the bottom of the cell and drown the egg or larva. There is clear evidence that this happens with some of the larger American species, so it has become an international recommendation to keep boxes in the normal orientation and not turn them upside down.

But inverting the brood of our Australian species does not hurt them. Their smaller cells mean that the contents adhere to the bottom and walls of the cells, so no damage is done. In fact, the brood cells of *Austroplebeia* species have their opening facing in all directions. For this species, at least, the viscosity and surface tension of the brood provisions allow it to remain in position.

11.5 Various brood separation scenarios

The separation of the brood will vary depending on the position of the advancing front, which is dynamic.

Now is a good time to revise what we know about brood rearing by stingless bees. Let's first consider *T. carbonaria* bees, which are the most common case. The advancing front of new cells is constantly being built upwards. The advancing front can be identified by the presence of newly constructed empty cells waiting to be provisioned with food, loaded with an egg and closed. As the building continues, the eggs hatch, develop through larval and pupal stages, and emerge as adults. The oldest comb, where the adults are emerging, disappears, creating a space (Figure 4-39, Figure 11-5).

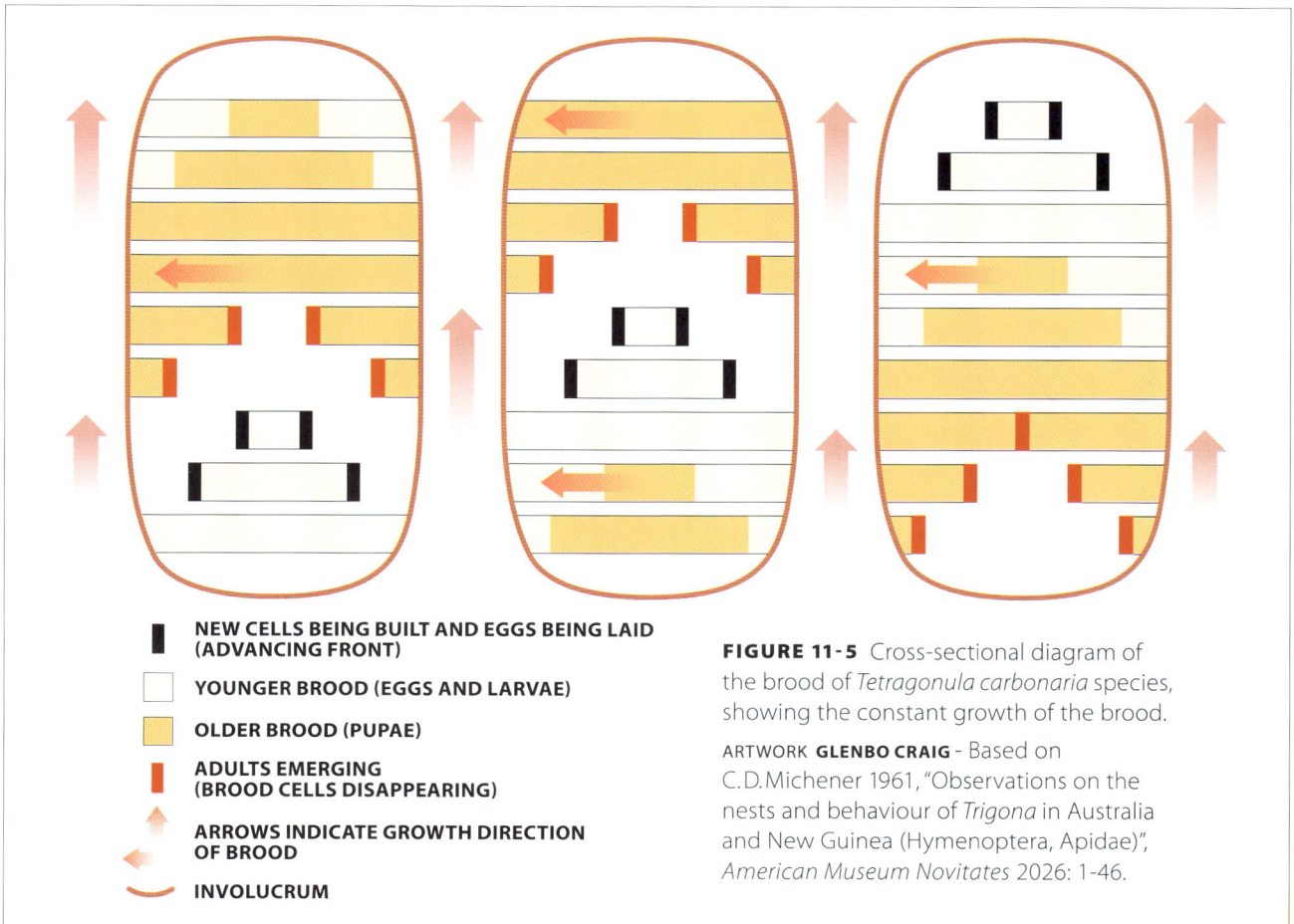

■ NEW CELLS BEING BUILT AND EGGS BEING LAID (ADVANCING FRONT)

☐ YOUNGER BROOD (EGGS AND LARVAE)

▨ OLDER BROOD (PUPAE)

▮ ADULTS EMERGING (BROOD CELLS DISAPPEARING)

⬆ ARROWS INDICATE GROWTH DIRECTION OF BROOD

⌣ INVOLUCRUM

FIGURE 11-5 Cross-sectional diagram of the brood of *Tetragonula carbonaria* species, showing the constant growth of the brood.

ARTWORK **GLENBO CRAIG** - Based on C.D.Michener 1961, "Observations on the nests and behaviour of *Trigona* in Australia and New Guinea (Hymenoptera, Apidae)", *American Museum Novitates* 2026: 1-46.

11.5.1 BROOD SEPARATES AT THE ADVANCING FRONT

The easiest separation of the comb occurs when it divides through the advancing front, which is a natural zone of weakness (Figure 11-6).

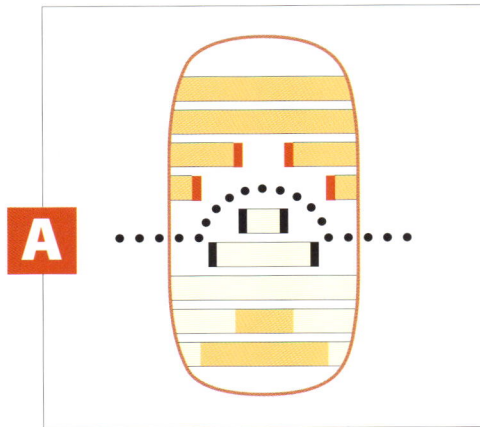

•••• INDICATES SEPARATION

FIGURE 11-6 In this division, the brood comb separated in Position A of the diagram, that is, at the advancing front.
IMAGE **JEFF WILLMER** – TAKEN AT ONE OF MY FIRST SPLITS IN 1985

11.5.2 BROOD SEPARATES AT THE LARVAL / PUPAL COMB

If the separation occurs through the larval or pupal comb, a little intervention may be needed. Remember that the comb is not actually separate sheets, but a spiral, so that each sheet joins the one above and the one below (Figure 11-7). It may be necessary to tear or cut the sheets to separate the layers with a knife (Figure 11-8). Sometimes it is also necessary to cut the involucrum and some food pots.

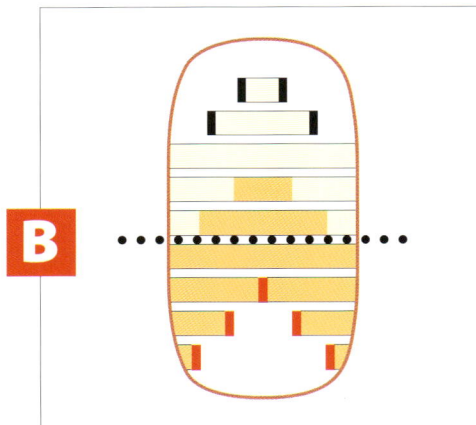

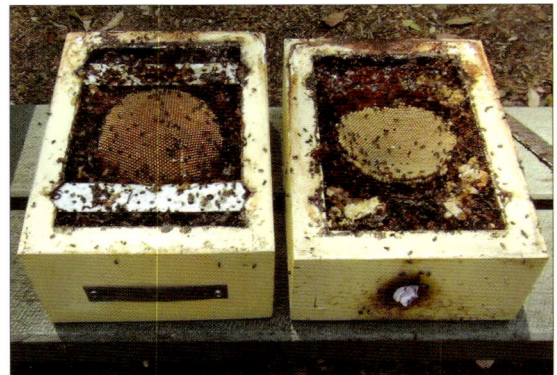

FIGURE 11-7 In this division, the brood comb was in Position B of the diagram, that is, through the larval comb.

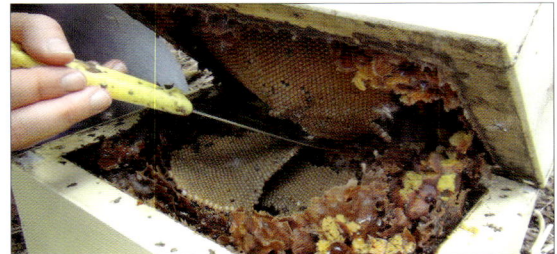

FIGURE 11-8 The brood comb being torn (*LEFT*), or cut with a knife (*ABOVE*) to separate the layers.

11.5.3 BROOD SEPARATES AT THE PUPAL COMB

If the division is through the pupal comb where the adults are emerging, then a ring of pupal comb with the advancing front underneath is visible (Figure 11-9). No intervention is needed.

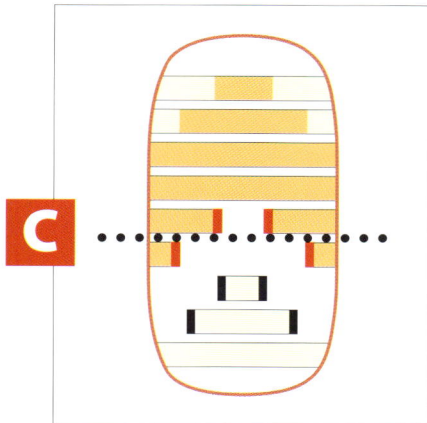

FIGURE 11-9 In this division, the brood comb was in Position C of the diagram, that is, through the pupal comb.

11.5.4 BROOD DOES NOT SEPARATE

Occasionally the brood does not separate into halves at all, but stays as a single unit in either the bottom or top hive section (Position D in Figure 11-10 and position E in Figure 11-12 respectively). One option in these cases is to close the hive and wait three to four weeks and split when the advancing front has moved to near the middle. Alternatively you can divide the brood and transfer one half into the other box. You may need to cut the involucrum with a knife in order to divide it.

If the brood chamber remains entirely in the bottom half, I suggest the following procedure. Turn the top section upside down. The involucrum, empty of brood, should form a concave space. Cut and remove the top half of the brood from the bottom box. Now place this top half in the concave space of the top box, keeping the brood in its original orientation (Figure 11-10). Close the hive by placing the new empty bottom section on top. This hive is now upside down, but the brood is not, and is supported from below. Leave the hive upside down until the next time you split it. Turning a hive upside down will not hurt it.

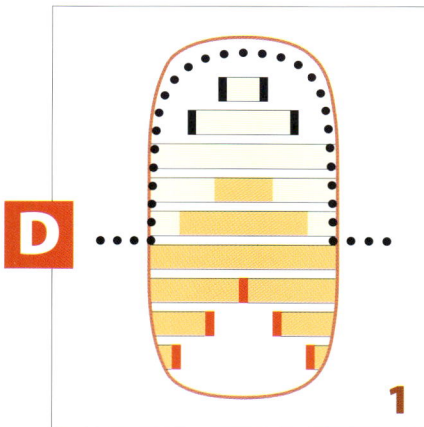

•••• INDICATES SEPARATION

FIGURE 11-10 1: This brood (of *Tetragonula carbonaria*) has not split but stayed as a single unit joined to the bottom section, that is, in Position D. **2:** Cut the brood. **3:** Separate the brood. **4:** Place the top half of the brood back in the top box, couple this box to the empty hive section and leave it upside down. IMAGES **JAMES DOREY**

Another option when the brood stays as a single unit in the bottom hive section also divides the brood but places the top half of the brood upside down in the empty new bottom hive section (Figure 11-11). This option works well when the brood is contained within and attached to its involucrum. The involucrum forms a stable base for the brood. Although it is inverted, the brood will survive and the bees will start building upon it.

FIGURE 11-11 This brood (of *Tetragonula carbonaria*) has not split but remains as a single unit in the bottom section. **1:** The top half of the brood was removed. **2:** The brood is inverted and placed in the bottom section of the empty new bottom section. **3:** This bottom hive section is joined with the occupied top section. IMAGE **JAMES DOREY**

If the brood chamber remains entirely in the top half, then you can carefully divide the brood in two and place the bottom half back into its space in the bottom section (Figure 11-12).

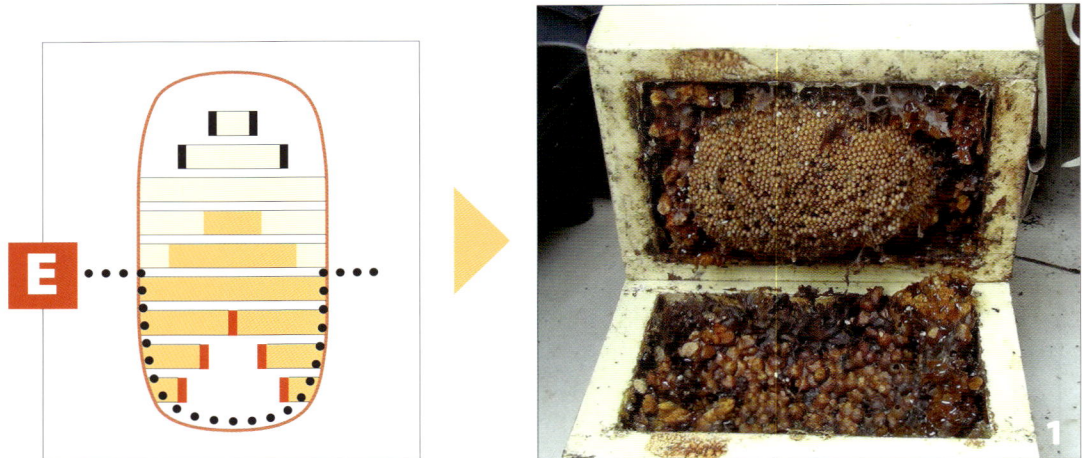

FIGURE 11-12 This brood (of *Tetragonula hockingsi*) has not split but stayed as a single unit in the top section. This sequence show the process of cutting the brood into two and dropping half into the bottom box, then proceeding with the hive division. IMAGES **DAVID MERRITT**

Note that some hives of some species, especially *T. hockingsi*, will occasionally form a swarm following the disturbance of a hive division (Figure 11-13). By the next day, they will have settled down and appear normal.

FIGURE 11-13 Some stingless bees, especially *Tetragonula hockingsi,* will at times form a swarm following the disturbance of a hive division.

11.5.5 RISK OF SLUMPING

There is a risk that the nest in the top box may slump into the bottom box following a division because it is no longer supported from below. This is a potentially fatal outcome, but is effectively prevented by split bars, or other anti-slump structures (Figure 11-1, Figure 9-3, and below).

If your hive is not fitted with split bars, you need to watch for potential slump and do something about it. Check that the top nest is securely connected to its hive box. If it appears to be dislodged and slumping, then you need to take action. I recommend two options: 1) while the hive sections are separated, apply tape around the front and back of the top section so that bands of tape sit in the same position as the split bars, or 2) continue with the division, but turn the hive upside down so that the new empty bottom section is on top. You can leave it like this or turn it back into its normal position after a few months when the top section will be stabilised with newly built connectives (Figure 4-6).

Wooden split bars

Relax!

If you open a hive and are not happy with the way it has separated and do not know how to proceed, do not worry or panic. Simply put it back together and try again later, when the advancing front will be in a new position and the division may proceed easily. Also do not worry if a few bees are sacrificed when dividing a hive. It is hard to totally eliminate casualties. The creation of a new colony of thousands justifies the loss of a few individual bees.

11.6 Dividing the hives of other species

The above description of splitting hives used mainly the example of the commonly kept *Tetragonula carbonaria*. The two other commonly kept species in Australia, *T. hockingsi* and *Austroplebeia australis*, can be divided in a similar manner. Probably most other species can be too. Below I explain in more detail how to divide hives of *T. hockingsi* and *A. australis*.

11.6.1 TETRAGONULA HOCKINGSI

Tetragonula hockingsi hives can also be split with great success using the standard OATH division. However, the brood architecture of this species is different from *T. carbonaria*. The less even "**semi-comb**" of *T. hockingsi* may not separate easily, which can complicate the operation. If the separation occurs at the advancing front (Figure 11-14), then it will proceed just as easily as *T. carbonaria*. The advancing front of a *T. hockingsi* colony is not as obvious as that of a *T. carbonaria*, but is still readily identified by the presence of newly constructed empty cells and cell contents that contain a lot of brood food and a small egg.

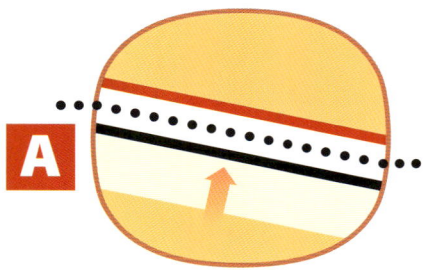

FIGURE 11-14 In this division, the brood comb separated in Position A of the diagram, that is, at the advancing front. Note the open cells that indicate the advancing front.

DIAGRAM BASED ON C.D. MICHENER 1961

If the *T. hockingsi* brood does not split at the advancing front, then it becomes a little more challenging. In the case of *T. carbonaria*, the flat layers of comb separate easily at any level and so the brood can be separated wherever the beekeeper chooses. However, the semi-comb brood of *T. hockingsi* will not separate easily and may need to be cut. It can be cut either in the horizontal or the vertical plane (Figure 11-15).

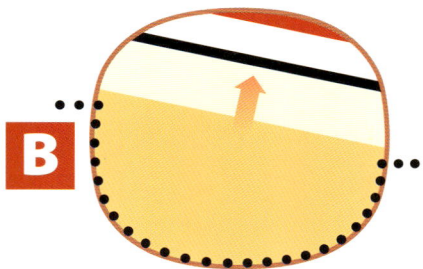

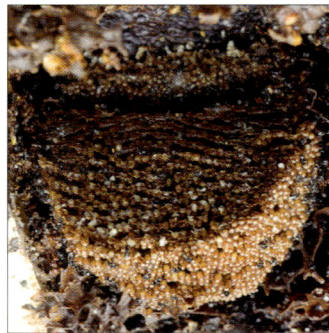

FIGURE 11-15 In this division of a *Tetragonula hockingsi* hive, the advancing front is near the top and so the brood has not divided evenly but all stayed in the top section, that is, in Position B of the diagram. As a result, the beekeeper has cut the brood in two vertically (Position C) and dropped one half into the bottom section.

INDICATES CUTTING

11.6.2 *AUSTROPLEBEIA AUSTRALIS*

Hives of *A. australis* can be effectively divided using the OATH split method (Figure 11-16). This species typically forms smaller nest volumes and so mini OATHS are recommended. (Read more about this in Chapter 9, "Constructing hive boxes for stingless bees".)

FIGURE 11-16 Splitting an *Austroplebeia australis* colony using the OATH method. **1:** The left hive has an observation window through which you can see that it is full; an empty three-section Honey OATH is ready on the right. **2:** The left hive has been opened and the brood split in two; the empty OATH has been opened in readiness. **3:** The top of the full hive is being placed on the empty bottom and the empty top section will be placed on the full bottom. IMAGES **CLAUDIA RASCHE**

Species of *Austroplebeia* show a cluster arrangement of brood (Figure 11-17). In addition, the cells of *Austroplebeia* do not always open upwards. This results in a different appearance when dividing. However, the brood typically separates into two halves, which will develop into two separate, independent colonies.

FIGURE 11-17

Cross-sectional diagram of the brood of *Austroplebeia australis* species, showing the constant growth of the brood from the centre outwards.

BASED ON **C.D. MICHENER** 1961

Image shows the hollow forming in the centre of the brood, as the adults emerge from cells in this space.

IMAGE **JAMES DOREY**

▮	**EGGS BEING LAID**
▯	**YOUNG BROOD**
🟧	**OLDER BROOD**
▮	**EMERGING ADULTS**
↑←	**ARROWS INDICATE GROWTH DIRECTION OF BROOD**
⌣	**INVOLUCRUM**
•••••	**INDICATES SEPARATION**

11.7 How does a colony re-queen itself?

When you divide a hive, one half will be queenless. But, even without dividing, every colony of bees becomes queenless when the mated queen dies, approximately annually. It is crucial to the survival of the colony that the queen is replaced as soon as possible. (Read more about the queen's role in Chapter 3 "Understanding the highly social bees".)

Our experience with Australian species is that you rarely need to intervene, because colonies normally re-queen themselves. In particular, Australian species of *Tetragonula* produce abundant virgin queens throughout the year (see *T. carbonaria* in Figure 3-10 and *T. hockingsi* in Figure 11-18). It is more difficult to observe in *Austroplebeia*, but they also seem to produce queen cells consistently (Figure 3-12). This may not be the case in the American tropics, where many species do not produce abundant queens.

There are four possible ways colonies can re-queen. These are listed below and described in detail in "Queen replacement" in Chapter 3:

1 a virgin queen present in the colony is recruited to the top job
2 a queen cell in the colony emerges and that virgin is selected to take over
3 an emergency queen cell is made
4 a nearby colony contributes a queen.

The recent discovery of emergency queens in *Tetragonula carbonaria* means we now know that, following artificial hive splitting, workers can correct the colony's queenless condition by producing queen cells from ordinary worker cells that contain eggs or young larvae. This relieves the pressure on you to inspect and ensure that potential queens are present.

FIGURE 11-18 A queen cell on the edge of the comb, in a *Tetragonula hockingsi* hive, ready for provisioning and egg laying by the mated queen.

11.8 What can you do if a hive has not re-queened?

If you believe that your hive is queenless and not likely to re-queen from any of the above natural methods, then you can transfer brood with queen cells from another colony (Figure 11-19). This brood will be accepted by the colony. The brood will give the recipient hive extra workers too. This is regularly practised by Peter Clarke in Sydney, where failure to re-queen may be more of a problem.

The brood of stingless bees remains alive outside the nest for a surprisingly long time. The nest temperature varies more in stingless bees than honey bees. Consequently, stingless bee brood seems to be more tolerant of temperature fluctuations than honey bee brood, allowing the storage and transport of brood for several days for transfer between colonies.

FIGURE 11-19 Removing a ring of brood with queen cells from one hive for placement in a queenless hive.

After dividing a hive, one hive should remain in the original position, but you need to find a new home for the new hive. You can move one of the hives far away (more than 1 km, which is beyond the flight range of the bees) or you can leave it nearby (usually on the same property). Let's look at these two options in more detail.

If you move a hive far away, the foraging bees from that hive will have to learn the new position. One strategy is to leave the original full top and a new empty bottom at the original location and move the original full bottom and empty new top to the new location. The bees from the former half of the division then get an unfamiliar entrance but this is balanced by retaining the original position. (Anyone who has moved a hive while bees are foraging will know just how loyal the returning foragers are to the original position. They will land and cluster very close to this position even though the hive itself is not there.)

Then you can take the other half of the division, the original full bottom and the empty new top, to a new location. These bees have to learn the new location but they have the advantage of retaining the original entrance with all its familiarity. Another advantage of this strategy is that the half that you move is more stable for transport because the contents are in the bottom box.

Often people wish to keep both hives on the same property (Figure 11-20). There is no problem doing this. But how much do the two hives need to be separated? There are a variety of opinions about how to position the new hive. Those strategies are largely about ensuring that the two hives each get a good number of forager bees.

BUT HERE'S THE THING: IT DOES NOT MATTER WHERE YOU PUT THE NEW HIVE!

Why? Because the foraging bees that you see leaving the hive and returning with pollen, nectar and resin do not do house duties. They are the oldest bees in the colony and have graduated from house duties, never to return to them. The majority of workers in a hive are house bees that are responsible for the duties of cleaning up after the split, sealing any gaps in the new hive walls, constructing the entrance tube, defending the entrance against attack by natural enemies, sealing up any broken pots, cleaning up any spilled honey and continuing the task of rearing brood. These are the jobs that urgently need to be done in the days after a division.

As long as the hive has some stored food, the hive does not urgently need the foragers. If you move a hive straight to another part of your property, regardless of how far away, most of the foragers will return to the original location. The hive in the new position may appear very quiet for a while, but do not fear, the house bees will not desert it, they are busy inside. Over the next few weeks the new foraging bees that graduate to this duty will bond to the new position and will start to bring back resources. So the most important factor in deciding position is ensuring a good micro-environment, not how far the position is from the original hive.

FIGURE 11-20 Three hives of stingless bees positioned closely in a front yard.

IMAGE
TOBIAS SMITH

12 Honey from stingless bee hives

Honey pots in a *Tetragonula* hive. IMAGE **GLENBO CRAIG**

Stingless bees collect nectar from flowers and process it into honey for later consumption.

The flavour of stingless bee honey depends on the source of the nectar used to create it, on the enzymes the bees add, and on the propolis pots used to store the honey.

Honey is about 80% sugar and 20% water, but the tiny percentage of remaining trace ingredients still influences **honey flavour and quality.**

Honey of Australia's most commonly kept stingless bee (*Tetragonula carbonaria*) has excellent **antimicrobial activity.**

Honey can be easily **harvested** from hives of the correct design (those that include a honey section). The thin honey is released by piercing the honey pots and captured in a tray.

A typical stingless bee hive will only yield about 1 kg of honey per year, but the **flavour is unique and exquisite.**

So you've saved a colony or two of native stingless bees from certain death at the hands of human-wrought destruction. You've set them up nicely in cosy hives and divided them to create new hives. You are enjoying their antics and maybe the benefits of their pollination services.

But there's more! You can harvest tangy, aromatic **sugarbag honey** from these little bees without harming the colony. And you can find out how to do this right here in this chapter. You might even find a recipe or two. So have a read and get stuck into it. You'll find the experience rewarding and delicious, albeit a little sticky! It's time to collect the rent.

Which species and where?

This section explains how to collect honey from Australian stingless bees, particularly *Tetragonula carbonaria* and *T. hockingsi*, the common east coast stingless bees, but much of it applies to other stingless bee species from other parts of Australia and the world. Some of the advice may have to be modified for *Austroplebeia* species, because these make smaller nests, use propolis of a different composition and produce honey of a different taste and properties.

Honey should not be extracted from colonies living at the cooler end of their range as they may need it to get them through the next winter. For example, it's a tough call to expect to collect much honey from *T. carbonaria* south or west of Sydney.

"Honey" vs "sugarbag"

The honey of stingless bees does not fall under some official definitions of honey, which include only honey made by honey bees. Hence, it may be better to avoid the word "honey" completely when considering stingless bees. In Australia, the term **sugarbag** is a widely used and appropriate word. This descriptor is obviously of English origin, but its use for honey of stingless bees was popularised by Aboriginal peoples. "Sugarbag" may refer to honey only or a mix of honey and pollen. Often these foods are stored together in the nest and end up mixed in the dillybag or container after collection from a nest (Figure 12-1).

12.1 How stingless bees make and store honey

Bees process nectar and pollen to preserve it for later use. In the case of pollen, the processing may improve its digestibility by their young.

In the case of nectar, the conversion to honey helps to keep it in a safe, usable state. Conditions in a stingless bee nest are warm and humid, ideal for the growth of microorganisms. Some of these organisms may cause the spoilage of stored food. Just as humans make cheese out of milk

FIGURE 12-1 A nest of *Austroplebeia essingtoni* opened to show brood, pollen and honey stores.

to preserve it, so too bees process their food so that it can be kept for later consumption. Here's how they do it.

At the flower, nectar is swallowed into an organ called the **crop**, in which it is carried back to the nest. Upon arrival at the nest, the forager contracts its abdomen to regurgitate the nectar to one or more house bees. These house bees ingest the nectar and carry it in their crop to nectar storage pots, which are not fully sealed but have an opening at the top through which the bee enters to deposit its load.

If the intake of nectar is greater than current needs of the colony, the excess will be made into honey for storage and later use. The process of making honey from nectar is called **ripening**. Honey ripening involves dehydration and addition of enzymes. Microorganisms may play a role in the honey ripening process by producing their own enzymes.

Storage and ripening take place in honey pots made of propolis or wax (Figure 12-2). A certain number of honey pots are always open within a hive. Some of these pots are used for unloading incoming nectar. From these pots, workers dehydrate nectar by ingesting a droplet, then regurgitating it and holding it in their mandibles. They move their tongue rhythmically through the drop, fan their wings to create a current of air over the droplet and gradually ingest it again. After many repetitions, the sugar content of the honey is increased and it is deposited back in the honey storage pot. (In one experiment with Trinidad's large stingless bee, the sugar content of storage pots started at 50% and increased 2–4% per day, reaching a final concentration of 70–80%.) The honey pot is then sealed. A newly sealed honey pot may be used for long-term storage or it may soon be opened again so that the honey can be consumed. Stingless bees may struggle to dehydrate the nectar because of the poor air circulation in their nests. Stingless bee nests are more closed and poorly ventilated than honey bee nests. So it is more difficult for stingless bees than honey bees to achieve a low water content.

During processing, the honey also becomes more acidic. High acidity inhibits the growth of microorganisms that may spoil the honey. The acidity is mainly caused by the addition of an

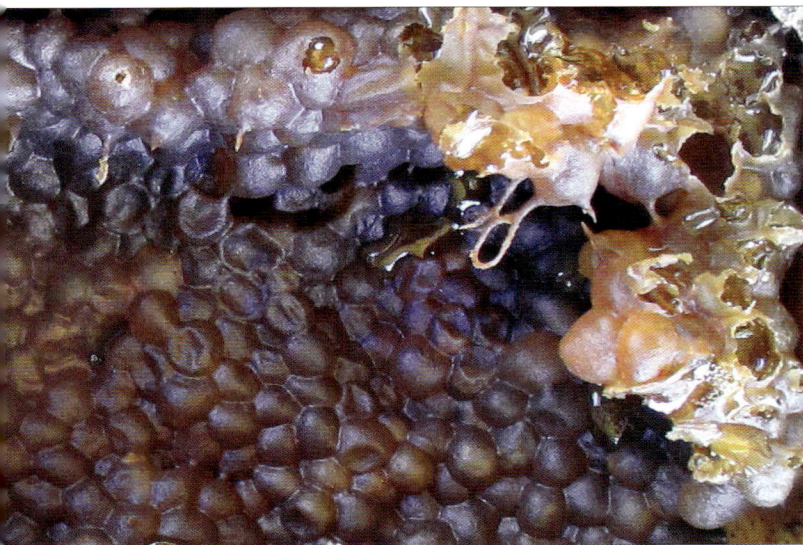

FIGURE 12-2 Honey pots of *Tetragonula carbonaria* (LEFT). IMAGE **MYFANY TURPIN**
Honey pots of *Austroplebeia essingtoni* (RIGHT).

enzyme, glucose-oxidase, which is produced in the salivary glands of bees. This enzyme converts glucose to gluconic acid (Figure 12-3).

The quantity of the reserves naturally stored in nests depends on the species and the conditions. *Tetragonula carbonaria* on Australia's subtropical east coast will normally store 1–2 kg of honey. *T. hockingsi* will store similar or even greater amounts. *Austroplebeia australis* and *A. cassiae* can also store large quantities of honey, but *A. essingtoni, T. mellipes, T. sapiens* and *T. clypearis* normally store less.

12.1.1 HOW STINGLESS BEES PROCESS AND STORE POLLEN

The foraging bee returning with a load of pollen carries the pollen to an open pollen storage pot by herself, without any transfer to a house bee. This may explain why pollen storage pots are usually found closer to the entrance of a hive than nectar storage pots (Figure 4-17). Usually, there are only a few pollen pots open. House bees quickly seal the open pots. This is probably an adaptation to combat nest parasites, which develop in pollen stores.

After depositing the pollen into the storage pots, house bees release a drop of clear liquid onto the pollen lumps and chew them with their mandibles to create a compact pollen mass. The pollen then ferments, giving the stored pollen a characteristic acid smell and sour taste. Later,

the preserved pollen is taken by the house bees to be deposited in brood cells as food for larvae. Only fermented pollen appears suitable for the rearing of brood.

How does this fermentation take place? Just as in the production of various foods by humans (e.g. beer, sauerkraut, cheese), microorganisms are involved. Let's look at this in more detail in the next section.

12.1.2 THE MICROBIOLOGY OF STORED FOOD

Microorganisms are believed to play a central role in the processing of the food of stingless bees. These bacteria and yeasts may be important in the conversion, fermentation or preservation of the food.

The microorganisms were originally probably parasites, feeding on and perhaps spoiling the food. But, after millions of years of evolution, they have formed mutually beneficial relationships with the bees. The bees provide a home and food for the microorganisms and, in turn, their food is preserved and made more digestible. These organisms thrive in the acid conditions and high sugar content of the bee provisions. Here, they play a fundamental role in preserving food by metabolic conversion. That is, they convert substances that are naturally in the food into new substances that inhibit the growth of organisms that may spoil the food. Furthermore, these bacteria apparently secrete chemicals such as antibiotics and fatty acids that inhibit competing microorganisms that could spoil food. In one revealing experiment, a Brazilian scientist used an antibiotic to eliminate the symbiotic bacteria in stored pollen of *Melipona quadrifasciata*, resulting in the death of the colony.

Other bee species are also known to host microorganisms in their food stores. For example, yeasts, moulds and bacteria belonging to the genus *Bacillus* are found in the stored pollen ("bee bread") of the honey bee. Bee bread differs biochemically and microbiologically from the pollen on the flower, and evidence suggests that microbes are responsible for the conversion of pollen to bee bread.

12.1.3 HONEY FLAVOUR

Honey bees store their honey in cells made of pure wax, which is a flavourless and odourless substance. Similarly, the honey of *Austroplebeia* species such as *A. australis, A. cassiae* and *A. essingtoni* is stored in wax pots and is sweet and viscous.

But the honey of Australian *Tetragonula* species is stored in pots made of propolis (Figure 12-2).

These honeys are aromatic, with eucalyptus-like flavours and aromas. Perhaps the resins in the propolis flavour the honey?

The honey of *Tetragonula* species is also sour, which raises the question: What purpose would this serve for the bee colony? I suggest that the resins and the acidity confer greater resistance to spoilage microbes, so the bees do not have to expend as much effort reducing the water content (water can encourage fermentation and spoilage).

Botanical sources (i.e. the types of plants in which bees forage) affect the flavour of honey bee honey. This is likely to also be true for stingless bee honey but, because stingless bee honey is made from nectar collected over a long period, it is more likely to be a mix of flower sources. Also, the resinous flavour of *Tetragonula* honey may mask the subtle flavours derived from different nectar sources.

12.2 Honey composition

Honey is essentially sugar and water; the remaining ingredients comprise less than 1%. Stingless bee honey is characterised by having a higher water content and higher acidity than most honey bee honey. The low water content of honey means that it draws water from its surrounding environment, so it can dehydrate the microbes that would otherwise grow and spoil the honey by fermenting it. An interesting question is how stingless bee honey remains preserved with this high moisture content. Honey bee honey with a moisture content of 25% is almost certain to ferment, which would downgrade the honey to "industrial" or "baking" grade. Yet stingless bee honey with a similar moisture content does not spoil. Other factors must prevent the spoilage of the honey, for example, high acidity and antibacterial substances from the resin used to make storage pots.

To learn more, we collected samples of honey from at least six hives of *Tetragonula carbonaria* in south-east Queensland and sent them to specialised laboratories around the world. Analysis of *T. carbonaria* honey shows that its composition is generally similar to that of honey produced by stingless bee species in other parts of the world (Table 12-1).

12.2.1 MOISTURE CONTENT

Moisture, or water, content is normally measured by a refractive index using a hand-held refractometer. Because it is so easily measured, it is reported more than other properties. The honey from our *Tetragonula carbonaria* proved to be an average of 26.5%, ranging from 25.3 to 27.5%.

This is very similar to the average moisture content for 152 samples of stingless bee honeys of many species from the American tropics, which was calculated at 27%. However, the moisture content of the American honey was far more variable, ranging from 19.9% to 41.9%. This is considerably higher than the honey produced by honey bees, which normally has a moisture content of less than 20%. This is why stingless bee honey is often less viscous (runnier) than normal honey.

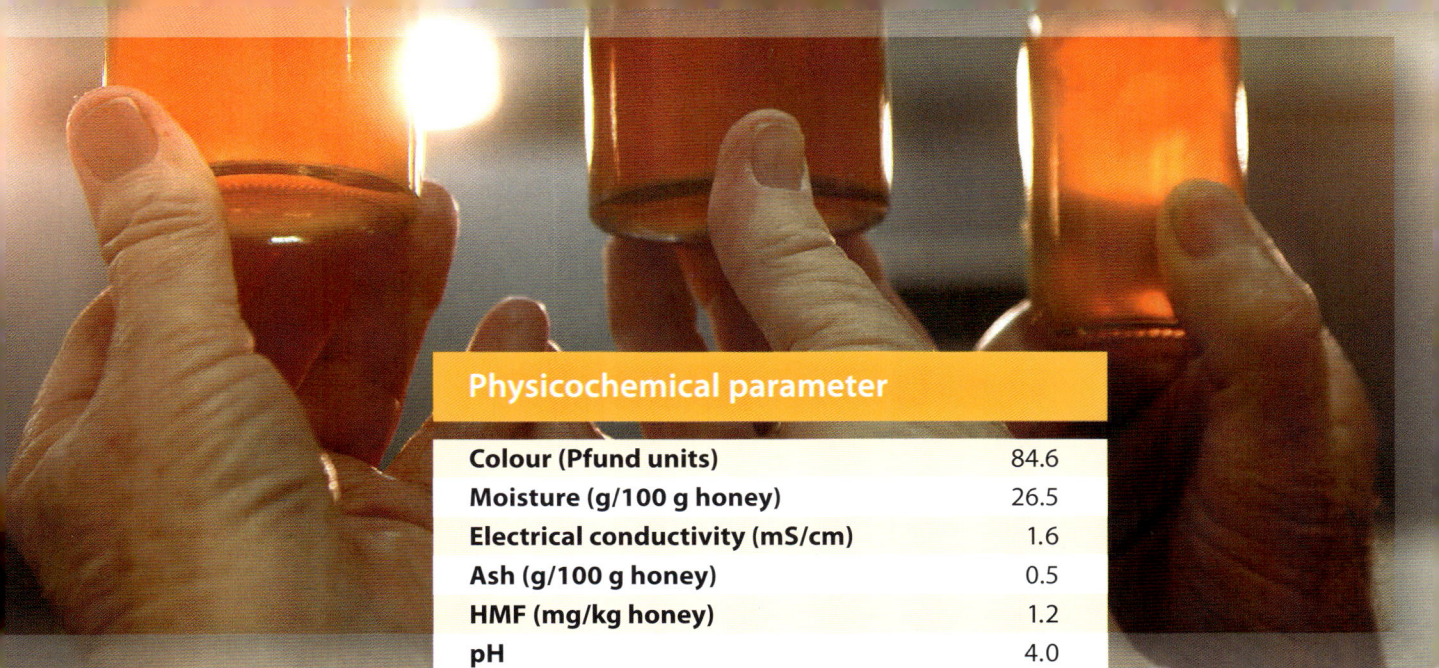

Physicochemical parameter	
Colour (Pfund units)	84.6
Moisture (g/100 g honey)	26.5
Electrical conductivity (mS/cm)	1.6
Ash (g/100 g honey)	0.5
HMF (mg/kg honey)	1.2
pH	4.0
Acidity (milli-equivalents/kg honey)	
Free acidity	124.2
Lactones	4.7
Total acidity	128.9
Nonaromatic organic acids	
D-gluconic acid (g/kg honey)	9.9
Citric acid (mg/kg honey)	228.7
Malic acid (mg/kg honey)	114.9
Nitrogen (mg/100 g honey)	202.3
Diastase (DN)	0.4
Invertase (IN)	5.7
Sugars (g/100g honey)	
Fructose	24.5
Glucose	17.5
Maltose	20.3
Sucrose	1.8
Fructose + glucose	42.0
Total sugars	64.6
Fructose/glucose ratio	1.4
Glucose/water ratio	0.7
Water activity (Aw)	0.7
Flavonoids (mg EQ/100 g honey)	10.0
Polyphenols (mg EGA/100 g honey)	55.7
Total antioxidant activity	
(µM Trolox equivalents)	233.9
Radical scavenging effect	
(% ascorbic acid equivalent)	48.0

TABLE 12-1

Summary of the physical and chemical analysis of *Tetragonula carbonaria* honey.

IMAGE
GLENBO CRAIG

As a result, stingless bee honey usually fails to meet Australian and international standards for the maximum water content permitted in honey bee honey. (Read more about this in "Some legal matters concerning honey" later in this chapter.)

An exception is *Austroplebeia* species, which make honey with a lower water content. These bees use more wax and little resin in their storage pots and their honey does not taste as acidic, so perhaps they rely on the lower water content to preserve their honey.

12.2.2 SUGAR

Sugars are the major components of honey. The composition of the sugars depends on the sugars present in the nectar and also on the sugars' transformation by the enzymes secreted by bees.

Honey from honey bees is dominated by the monosaccharides fructose and glucose (called "reducing sugars"). In Australia, the minimum permitted level of reducing sugars in honey bee honey is 60%, while 65% is specified in some overseas codes.

Small amounts of disaccharides and trisaccharides are also present in honey bee honey. In particular, the disaccharide sucrose can be present in quantities up to 5%. Monosaccharides are sugars composed of one sugar unit, while disaccharides have two sugar units. For example, sucrose is a disaccharide consisting of one glucose and one fructose unit, and maltose is a disaccharide consisting of two glucose units (Figure 12-3).

Stingless bee honey can be similarly dominated by fructose and glucose. For example, in a review of stingless bee honeys of many species from the American tropics, the average reducing sugars content was 66% and ranged from 58% to 76%. Very few of these honeys were below the 60% minimum required by the quality standards for honey bee honey; however, *Tetragonula carbonaria* had only a little over 40% reducing sugars. This was due to the presence of the disaccharide maltose, or a sugar very similar to it. The exact nature of this mystery sugar is yet to be confirmed. Its presence could be due either to the different origins of the nectar or to its enzymatic transformation by the bees.

12.2.3 ENZYMES: DIASTASE, INVERTASE AND GLUCOSE-OXIDASE

Enzymes are biological catalysts, or chemicals that speed up chemical reactions. They originate in the hypopharyngeal and salivary glands of bees (Figure 1-10).

Diastase

Diastase is one of the enzymes commonly found in honey. Diastase, the common name for α-amylase, is responsible for the digestion of starch into sugars. Diastase is probably important in the digestion of pollen. The diastase number is determined by the rate of starch hydrolysed. It usually varies from 9 to 32. The minimum acceptable level according to the Codex Alimentarius (an international food code) is 8, but stingless bee honey normally contains less diastase than this. The value in the *Tetragonula carbonaria* samples was only 0.4 (Table 12-1).

Invertase

Invertase is the enzyme added to the nectar by the bee that converts the nectar's sucrose to glucose and fructose, a central step in the ripening of nectar to honey. Why so? Because these two monosaccharides can be dissolved in water to a much higher concentration than sucrose in its original form. This means the bees can make a thicker honey that can be stored for longer. Clever, aren't they?

Invertase activity is a useful freshness indicator. Although there is great natural variation, the invertase number should be greater than 10. In studies on American stingless bee species, the invertase numbers were all greater than 10. On the other hand, the value for *Tetragonula carbonaria* was 5.7. We do not yet understand the significance of this lower value.

Glucose-oxidase, gluconic acid and hydrogen peroxide

Glucose-oxidase converts glucose into gluconic acid and hydrogen peroxide. Gluconic acid lowers the pH of the honey. Hydrogen peroxide produces free oxygen and hydroxyl radicals that have an antibiotic effect (Figure 12-3). Hydrogen peroxide prevents spoilage of the nectar until it is sufficiently dehydrated to prevent microbial growth. Both gluconic acid and hydrogen peroxide occur in the honey of honey bees and also of stingless bees.

12.2.4 HYDROXYMETHYLFURFURAL

Hydroxymethylfurfural (HMF) is a major quality factor in honey. It is a measure of freshness and of overheating. In fresh honey, there is very little HMF, but it increases with storage, especially at the high temperatures experienced in warm countries. A maximum value of 40 mg/kg is set for European Union honey.

Most stingless bee honeys studied so far remain below this number, but no studies have been done on honey that has followed common market conditions of heat and storage. The value for *Tetragonula carbonaria* was 1.2.

12.2.5 ASH

Ash is what is left after a honey is ignited to remove all the organic components and water. It consists of minerals. A typical value in honey from honey bees is 0.2%. It was 0.5% in *Tetragonula carbonaria*.

12.2.6 ACIDITY AND pH

Stingless bee honeys are acidic or sour honeys. This is reflected in their high acidity (measured in milliequivalents/kg) and low pH. This high acidity is probably important in keeping these honeys unspoiled at high moisture contents. It may also partly explain the high antimicrobial activity.

Acidity, sometimes called free acidity, is an important quality criterion in all honeys, because honey fermentation causes an increase in acidity. The maximum acceptable acidity in honey bee honey is 40 or 50 meq/kg. Acidity is frequently much higher than this in stingless bee honey. The value for *Tetragonula carbonaria* typifies this at 124.

pH is not a characteristic of honey that is considered important in food standards, but it is sometimes measured. Stingless bee honey has a lower pH than honey bee honey. The average pH of honey bee honey is 3.9. Typical pH values for stingless bee honey are 3.3 to 4.0. The honey of *Tetragonula carbonaria* was pH 4.0. The low pH of honey is at least partly due to the presence of gluconic acid.

12.2.7 ELECTRICAL CONDUCTIVITY

The electrical conductivity of honey bee honey can be used to distinguish the honey's origin. The conductivity of honey bee honey from blossom should be less than 0.8 mS/cm. The high conductivity of *Tetragonula carbonaria* (1.6 mS/cm) is probably due to its high pollen content.

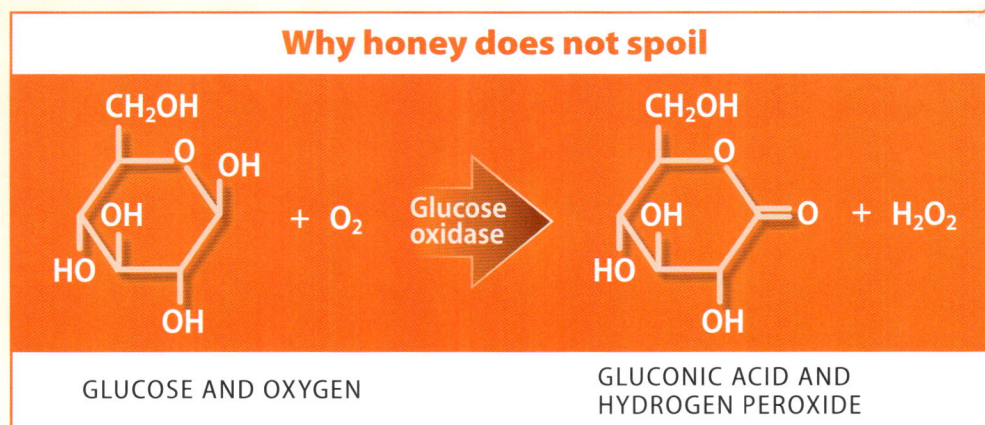

Why honey does not spoil

$$CH_2OH \cdots + O_2 \xrightarrow{\text{Glucose oxidase}} CH_2OH \cdots + H_2O_2$$

GLUCOSE AND OXYGEN → GLUCONIC ACID AND HYDROGEN PEROXIDE

FIGURE 12-3
Why honey does not go off.
ARTWORK
GLENBO CRAIG

Other than its low water content, honey is acidic, with a pH between 3 and 4. This acidity inhibits the growth of microbes. The acidity is created when enzymes from the bees' saliva convert glucose to gluconic acid. This chemical reaction also generates hydrogen peroxide, which also makes honey a hostile environment for microbial growth.

12.3 Medicinal properties of stingless bee honey

Stingless bee honeys are used by native peoples around the world as a medicine. (Read more about this in the feature "Indigenous peoples and stingless bees".) Their faith in the medical properties of stingless bee honey is now supported by scientific research.

12.3.1 ANTIBIOTIC ACTIVITY

The honey of *Tetragonula carbonaria* has excellent antimicrobial properties. This has been corroborated by studies at three Australian university laboratories that used a variety of techniques to analyse the effect of stingless bee honey on a range of pathogenic microbes.

Compared to manuka medicinal honey, stingless bee honey rates well. Manuka honey is harvested from hives where honey bees have collected nectar from manuka (*Leptospermum*) flowers. In the tests, two aspects of activity were tested: total and non-peroxide. Hydrogen peroxide is an antimicrobial agent present in nearly all honey, including that from stingless bees. But some honeys possess factors other than hydrogen peroxide. Manuka honey contains a powerful antimicrobial agent called methylglyoxal (MGO), which comes from the nectar of *Leptospermum* flowers. The results were positive; even when the activity offered by hydrogen peroxide was removed, the honey of *T. carbonaria* was very active (Figure 12-4). In addition, the longevity of the activity is reasonable, with a loss of only 15% at 28 weeks.

So what is in carbonaria honey that gives it its kick? Honey chemist Flavia Massaro showed that it does not contain MGO, the most active component of manuka honey, or other markers of *Leptospermum* nectars. But the honey does contain flavonoids that inhibit microbial growth. These flavonoids probably originate in the resin used by stingless bees to build their honey pots (Figure 12-5). (Read more about resin in "Building materials: wax, resin and propolis" in Chapter 4.) In contrast, honey bees use only wax to build the cells that store food, so their honey does not make contact with propolis.

The weight of evidence supports this theory, but there is another possible explanation. This is that "good" microbes in the honey of stingless bees produce substances that inhibit the growth of microorganisms that would spoil the honey. Either way, by overcoming the problem of food spoiling in the nest, stingless bees create a honey with potential medicinal properties.

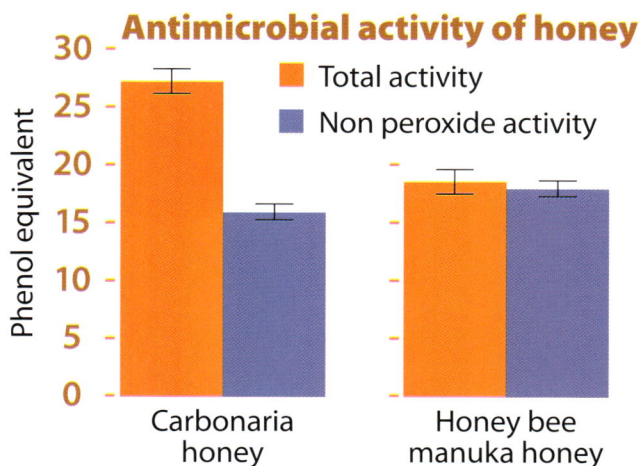

Antimicrobial activity: where is it from?

Honey comb of *Apis melifera*
Made of **WAX**

Honey pots of *Tetragonula carbonaria*
Made of **PROPOLIS**

FIGURE 12-5 The honey of honey bees is stored in comb made of wax *(ABOVE LEFT)*. The honey of stingless bee *Tetragonula carbonaria* is stored in pots made of propolis, which has antimicrobial properties that may infuse the honey *(ABOVE RIGHT)*.

IMAGES **GLENBO CRAIG** AND **KATINA HEARD**

Antimicrobial activity of honey

- Total activity
- Non peroxide activity

(y-axis: Phenol equivalent, 0 to 30)
(x-axis: Carbonaria honey, Honey bee manuka honey)

FIGURE 12-4 The honey of the stingless bee *Tetragonula carbonaria* has antimicrobial properties similar to the honey that honey bees make from manuka flowers.

12.3.2 HONEY AS AN ALLERGY REMEDY

Local raw honey from honey bees is sought by many allergy sufferers because its pollen impurities are thought to reduce people's sensitivity to hay fever. However, controlled studies have shown that honey bee honey is no more effective than placebos in alleviating allergies. This may be because most seasonal allergies are caused by the pollens of wind-pollinated plants, such as grasses, which bees generally do not collect. No studies have yet assessed whether sugarbag honey from stingless bees might be a more effective allergy remedy.

12.4 Harvesting honey from stingless bee hives

12.4.1 EXTRACTING HONEY FROM A HONEY OATH HIVE USING THE PIERCE AND DRAIN METHOD

Extracting honey from a hive with a dedicated honey section is simple: just pry, pierce, drain and strain. To use this method, you need a hive that has a section on top for storing the honey. (Read more about this in Chapter 9 "Constructing hive boxes for stingless bees".)

First, remove the honey section by loosening it with a hive tool or other lever and prying it open. Then invert the honey section so that the open side faces up, and pierce the honey pots. I use a tool that resembles a miniature bed of nails to do this (Figure 12-6). The depth of the honey hive can be up to 65 mm, so the nails should protrude at least that much from their mounting.

I then turn the hive over onto a plastic container to catch the tide of liquid gold that issues forth. Allow the honey to drain for at least five minutes. Ten minutes may be required, especially in cooler weather when the honey will not flow easily. Do not place the honey section back on the hive if it is still dripping heavily, because it may spill too much honey for the workers to clean it up quickly and may also attract natural enemies. You do not need the centrifugal forces generated by a commercial honey extractor to coax the honey out because sugarbag honey is runnier than that from honey bees.

After the honey section has fully drained, put it back on top of the hive. Strain any trapped bees out

of the honey and put them on the roof of the hive. You may wash or spray them lightly with water to wash off excess honey. Many will recover and make their way back into their hive. Back home in your kitchen, you can strain the honey again using a finer gauze to filter out any impurities.

The honey pots damaged during honey extraction will be quickly rebuilt. When the hive has regained weight, you can repeat the extraction procedure and you will find the honey pots rebuilt as good as new.

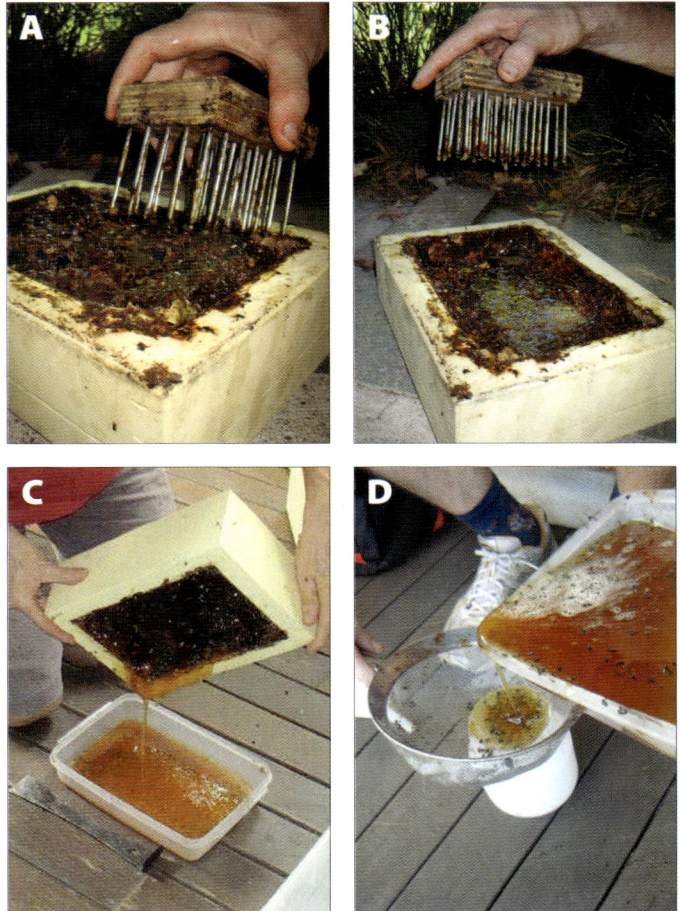

FIGURE 12-6
A Piercing the honey pots with the bed of nails.
B Making sure that the pots are pierced well and the honey is released.
C Draining the honey from the section into a container.
D Straining the honey.
IMAGES **KATINA HEARD**

The Honey OATH design allows the rapid extraction of honey with minimal impact on the hive. Studies show that bee mortality rates in these hives are not elevated by honey extraction. Some bees do die in the extraction process, but this is of little consequence to the colony. Certainly, this mortality is insignificant compared to the losses of thousands of bees from hives when they engage in natural fighting swarms.

A hive of 8.5 litres internal volume usually needs to contain approximately 5 kg of contents before it is likely to be full, including a full honey section. Honey can be extracted at any time of year using the Honey OATH design, which leaves sufficient stores of food in the brood boxes. An advantage of extracting in cold weather is that the bees move into the brood box to keep warm or to help to warm the brood, leaving the honey relatively free of bees.

12.4.2 ALTERNATIVE EXTRACTION TECHNIQUES

Individual beekeepers have developed their own honey extraction methods for stingless bee hives. For example, Chris Fuller and Tony Goodrich place plastic containers in the honey section of the hive during the establishment phase, so that it can be easily removed later to produce an instant container of honey (Figure 12-7).

FIGURE 12-7 Chris Fuller's collection system full of honey *(LEFT)*. IMAGE **CHRIS FULLER**
Tony Goodrich's honey collection system with four empty containers *(RIGHT)*.

Another method is to use a syringe attached to a suction pump to extract honey from pots. This method is effective but time-consuming. It is used in Asia and South America, where they have stingless bees that make larger honey pots. In Australia, where labour costs are high, this technique is unlikely to gain acceptance. But, if you have the time, it will result in a pure honey with no pollen.

12.4.3 PRODUCTIVITY OF HIVES

I started testing and fine-tuning the Honey OATH design in 1998. By 2001, I was satisfied that it was working well and started increasing hive numbers. By 2004, I had built up my numbers to approximately 50 hives of *T. carbonaria* and *T. hockingsi* across a number of suburban locations in south-east Queensland. For four years, I measured the productivity of these hives. They yielded a total of about 50 kg of honey per year, with an annual average production of 850 g per hive. I extracted the honey every 8 months.

This provides a rough guide to what may be expected from a colony of stingless bees, but honey production will vary with location, bee species and hive design. Location is very important: hives in Sydney may not produce any harvestable honey at all, while hives in northern Australia may produce higher yields. Also, suburban areas are more productive than rural areas or forest. Productivity of hives of different bee species will vary, especially in relation to locality. Hopefully, future innovations in hive design and management will promote greater productivity.

We should not expect stingless bee hives to approach the productivity of honey bees. Honey bees weigh 100 mg, about 25 times more than the 4 mg of a *T. carbonaria*; no wonder they can fly a lot farther and carry a lot more pollen and nectar. The population of honey bee hives is also higher than colonies of *T. carbonaria*, so it is not surprising that a hive of honey bees can produce 50 kg of honey in a year — about 50 times more honey than a stingless bee hive.

I gave up producing honey in 2010 to concentrate on producing hives. The only honey extraction I do now is to demonstrate the process at workshops. The process is a star attraction there, and the end result, sugarbag drizzled on ice-cream, is loved by the attendees (Figure 12-8).

FIGURE 12-8 Freshly extracted sugarbag honey is a rare and splendid treat, here served on ice cream.
IMAGE **GLENBO CRAIG**

12.4.4 PROBLEMS AND SOLUTIONS IN HONEY PRODUCTION

Honey section is sealed off

Sometimes stingless bees are overwhelmed by the extra space provided by a honey hive. Especially if a colony is weak, workers will seal it off with a propolis wall. If this occurs, just poke a small hole through to remind the workers that the space is there and they will start using it when they need extra room.

Honey section is empty

Many a disappointed beekeeper has opened a hive that recorded a good weight, only to find nought in the honey section (Figure 12-9). If this happens, the bees may have chosen to completely fill the available space in the brood box with stored food before starting in the honey box. The presence of propolis structures and wall finishes is a good sign that the space is being prepared for use. Close the hive and wait.

FIGURE 12-9
An empty honey section, separated from the section below. The bees have applied a coat of propolis to much of the surface in readiness to occupy it.
IMAGE **KATINA HEARD**

Pollen is stored in the honey section

Sometimes a large amount of pollen is stored in the honey section (Figure 12-10) and it becomes a challenge to extract the honey without getting lots of pollen mixed with it.

In this case, the honey is a very different product from pure honey. It has a complex and sour flavour. It is initially very cloudy and, later, the pollen tends to float to the top and form a layer or crust that not everyone will find attractive. This blend of pollen and honey is prone to fermentation; perhaps the populations of microorganisms in the pollen explode with access to the sugar and act on the honey to ferment it.

The result is an extraordinary food, which I relish. There may be a market for this complex blend of honey and pollen; after all it most closely resembles the tucker extracted from hives and eaten by Aboriginal people. But it should be distinguished from pure honey. Not everyone will like it and the reputation of sugarbag honey may be compromised if consumers get a batch of sour stuff with a frothy coating! Also, work is needed on how to get it to market in good condition, as its flavour does degrade quickly.

Pollen

FIGURE 12-10 A large amount of pollen has been stored in this honey section.
IMAGE **KATINA HEARD**

Many will want pure honey and not wish to extract pollen. So what do you do with it? It has to be removed or the space is not available for bees to store future honey production. I sometimes use it for my own consumption and that of my friends (my brother can eat half a kilogram of it by himself!). On other occasions, I keep it and allow it to stand. The pollen rises to the top and can either be skimmed off or the honey drained from below (Figure 12-11). Alternatively, I cut it out and feed it to another hive in need of food. If you do this, be careful not to break the pots too much.

Pollen floating on top

Pure honey

FIGURE 12-11 The bottle on the left is pure honey, while the bottle on the right contains honey that had a lot of pollen mixed in. After 24 days, the pollen has risen to the top so the honey can be drained from below.

Propolis or resin builds up in the honey section

This is only a problem if you are a collector of honey only and not propolis. Turn it into an opportunity and become a propolis producer. (Read more about this in the feature "Propolis from stingless bee hives".)

The honey foams when it is extracted

The honey sometimes foams when extracted. This foam rises to the top like the head on a glass of Guinness beer. However, unlike the foam on a Guinness, it may stubbornly persist for days after extraction. This is probably caused by a lot of unripe honey in the honey section. I recommend that you eat this honey quickly and do not store it for too long as it may quickly deteriorate.

12.5 Storage of honey

There is no doubt that sugarbag is at its most delicious when freshly removed from the hive. The flavour slowly deteriorates with time and the honey can ferment. When a container of older sugarbag honey is opened, a slight hiss may be heard as the pressurised gas, produced by fermentation, escapes. This may not be a serious issue. The slight fermentation of the honey may be considered one of its characteristics, just like the smelliness of a cheese that is also still "alive". But this is likely to be resisted by consumers who associate a gas build-up with spoilage, so ways to prevent sugarbag from fermenting or otherwise changing too much from its extracted state follow.

12.5.1 REFRIGERATION

Keeping the honey in a domestic refrigerator is a useful way of prolonging its shelf life. Scott Middlebrook recommends freezing it and feels that this has no negative impact on the flavour.

Honey can also be kept at room temperature for some time, but probably lasts better in the dark. Light may destroy valuable enzymes, especially glucose oxidase that produces the antibiotic agent hydrogen peroxide.

12.5.2 CAREFUL EXTRACTION

It seems that the presence of excessive amounts of pollen in sugarbag can hasten its fermentation. This is probably because the microbes that live in the pollen increase in abundance when given access to extra sugar. One way to minimise this is to take care not to adulterate the sugarbag with pollen during the extraction process.

12.5.3 GENERAL HYGIENE

Honey is self-preserving but this does not mean that hygiene is not required. Always use clean utensils. My utensils are stained with propolis, which makes them *look* shabby, but I strive to keep them as clean as possible, along with the work area and containers for storage or sale.

12.5.4 PASTEURISATION

Pasteurisation, or heat treatment, is a process of heating a liquid food to a certain temperature for a certain time. This simple but effective process kills most of the harmful organisms in foods and makes them safer to eat and last longer.

For example, the safety and longevity of milk is increased by pasteurisation.

It is true, however, that milk is a non-durable product — unlike honey, which has remarkable powers of self-preservation. So why bother to pasteurise honey? Well, even this product, famous for its antibiotic properties, can ferment, so pasteurisation may be useful in some situations (for example, if you wish to send it on a long overseas journey).

Brazilian Paulo Nogueira-Neto, in his wonderful book on keeping stingless bees (unfortunately only available in Portuguese), teaches us how to pasteurise honey by placing it in a pot of water, heating the water until the honey reaches 72°C, and holding it there for 15 seconds. The honey should be stirred during the heating process.

The heat can then be turned off and the honey allowed to slowly cool while remaining in the water. The cap of the honey container must also be removed and sterilised in the heated water. A thermometer is required.

An alternative treatment is 65°C for 30 minutes, but the longer duration is hard to achieve without special equipment. These treatments are said to have minimal impact on the flavour of the honey.

12.6 Selling honey

We estimate that the total current annual sugarbag production in Australia is extremely small, around 250 kg. But there is potential for rapid growth. The market price of the honey is currently about AU$100 per kg wholesale, or $200 per kg retail. This high price reflects the rarity of the product and the expense of producing it. A future challenge will be to maintain or increase the price while production is growing. This will require skilful marketing. Honey producer Scott Middlebrook suggests that producers need a honey cooperative. A cooperative could produce standard labels, provide advertising and be a form of information-sharing. The first sugarbag competition was recently held in Gympie, Queensland (Figure 12-12).

Restaurants that promote native or novel foods are starting to express an interest in sugarbag honey. Other potential sales points are tourist centres, airport shops, gift shops, health food shops and native plant nurseries. Demand is expected to grow rapidly as awareness increases. If you are interested in selling sugarbag, it may be helpful to display a living bee hive where possible so that customers appreciate exactly what the product is.

FIGURE 12-12 The entries in Australia's first official sugarbag competition, held in Gympie in March 2014.
IMAGE **GLENBO CRAIG**

12.6.1 SOME LEGAL MATTERS CONCERNING HONEY

Sugarbag is a food product, and so is governed by the Australia New Zealand Food Standards Code, which defines honey and prescribes its composition. However, as the code's provisions have been devised with reference only to honey bee honey, some issues need to be resolved if sugarbag honey is to be appropriately regulated.

For example, Standard 2.8.2 stipulates that honey should contain: (a) no less than 60% reducing sugars and (b) no more than 21% moisture. Sugarbag does not comply with these standards because it normally contains about 25% moisture.

International food standards also set a maximum permissible acidity because acidity is an indicator of honey fermentation. The maximum acceptable acidity in honey bee honey is 40 or 50 meq/kg. Acidity of stingless bee honey is frequently much higher than this. This is an international problem: prominent honey scientist Patricia Vit has also noted that honeys from Venezuelan stingless bees do not fulfil the quality requirements of the international Codex Alimentarius. Because

IMAGE **GLENBO CRAIG**

is pure honey, an ingredient label may not be mandatory.

Warning and advisory declarations

Under clause 4 of Standard 1.2.3 Mandatory Warning and Advisory Statements and Declarations, the presence of bee pollen in a food must be declared whenever present as an ingredient or additive. The definition of ingredient in Standard 1.2.4 is "any substance used in the preparation, manufacture or handling of a food". As bee pollen is naturally present in honey, it may not need to be declared.

Nutrition information panels

A generic nutrition information panel can be used for all varieties of honey. To provide the necessary information on a nutrition information panel, food composition data is required. Food composition data can be obtained either from food composition tables or databases, laboratory analysis, or the nutrition panel calculator provided by Food Standards Australia New Zealand, which has been designed to assist manufacturers. The calculator is available free-of-charge at www.anzfa.gov.au. An example of a nutrition information panel for stingless bee honey appears below.

You should seek legal advice regarding food standards and labelling requirements before selling sugarbag honey.

of these discrepancies, there could be legal concerns if some party challenged the lack of legal compliance of stingless bee honey. To minimise this likelihood, the industry should make all efforts to abide by regulations.

Labelling requirements

Part 1.2 of the Australia New Zealand Food Standards Code sets out the information that must be provided on foods that are required to bear a label. The label on a package of pure honey for retail sale must include the following general information:

• the prescribed name of the product (i.e. honey)
• the lot identification
• the name and business address of the supplier
• date marking
• nutrition labelling.

The label must also include a statement about the country of origin of the product. If the product

NUTRITION INFORMATION
Serving size: 7 g

	Quantity per serve	Quantity per 100 g
Energy	88kJ	1258kJ
Protein	0.02g	0.2g
Fat total	0g	0g
Carbohydrate	5g	74g
- Sugars	5g	74g

Ingredients 100% pure honey from Australian native stingless bees.

FIGURE 12-13 A possible nutrition panel for containers of sugarbag honey.

FEATURE

Recipes using sugarbag honey

Sugarbag honey is gaining quite a reputation as a gourmet sensation. As celebrity chef Kylie Kwong wrote to me:

Oh my God!!!!

How **stunning** is your honey!!!!! Nectar of the Gods. Like a superb dessert wine.

Oh my goodness, it is definitely **liquid gold!**

We were blown away by it, the texture, that lemony zing, the eucalyptus, oh God it's good!

I am serving it at the moment: 'Deep-Fried Silken Tofu with Roasted, Caramelised Organic Tomatoes, Salt Bush, Shiro Shoyu & Naturally Fermented Sugarbag Honey'– we serve them with your honey on the side, so they can experience it in **all its glory!**

With comments like this from such a well-known chef, the future looks bright for this product.

As a drizzle

Sugarbag goes very well drizzled over vanilla ice-cream. Its tanginess and spicy aroma complement the ice-cream's sweet smoothness. One day I expect that ice-cream makers will add it to their repertoire. Native food pioneer Vic Cherikoff recommends sugarbag drizzle as a topping on any dessert. "Also excellent for tarts, fruit compotes or to flavour any cold dish. The taste is like a blend of quality honey and a mellow port." Goat's cheese, a delight when served with bread or biscuits as a starter, can be enhanced by drizzling with honey. If that honey is sugarbag, then it becomes an even more interesting and exquisite treat.

Sugarbag and whipped yogurt cocktail

Glenbo's all-time favourite fare:

Hand whip (aerate) natural plain Greek-style yogurt with a few drops of sesame oil, until smooth and creamy.

Top with fingerlime, then lightly swirl sugarbag honey into the mix. Serve with anticipation!

IMAGE **GLENBO CRAIG**

Tim's sugarbag and macadamia pie

You can't get more Aussie than this exquisite treat that uses nuts native to Australian rainforests combined with sugarbag. It's my adaptation of pecan and maple syrup pie and it's sooooo good.

Ingredients

PASTRY
125 g butter
¼ teaspoon baking powder
220 g plain flour (1½ cups)
4 tablespoons of water

FILLING
½ cup brown sugar
½ cup sugarbag (=125 g)
  (keep a little to drizzle over when serving)
50 g butter (melted and cooled)
3 eggs
1 cup macadamia nuts
1 dessertspoon of lemon juice
  (or grated rind of one orange)

Method

- Add flour and baking powder to a bowl.
- Chop butter into small pieces and rub into flour with the fingers until it resembles breadcrumbs.
- Add 4 tablespoons of water and mix into dough.
- Knead dough until it is smooth, adding a little more water if necessary. Wrap in plastic film and chill in refrigerator for 15 minutes.
- Meanwhile, preheat oven to 180°C.
- Roll out pastry between two sheets of plastic film to fit into a pie dish (approx. 250 mm diameter round or 220 mm x 220 mm square).
- Line dish with pastry.
- Bake pastry for 15 minutes.
- If the macadamias are raw, toast them in the oven for about 5 minutes or until golden, being careful not to overdo them.
- Make the filling by beating together all the ingredients.
- Remove pastry from oven and reduce oven temperature to 150°C.
- Pour filling onto pre-baked pastry. Bake for about 30 minutes or until the mixture is just firm.

Serving suggestion

Serve warm or cold with ice-cream or cream and extra sugarbag drizzled on top.

Tasmanian salmon seared with native pepper, served with kipflers and snow peas and a sugarbag and finger lime dressing

Stephane Bremont, formerly of Tukka Restaurant in Brisbane, pioneered the use of sugarbag in fine cuisine. Here is one of his recipes (with some simplifications).

Ingredients

800 g salmon
300 g kipfler potatoes
80 g snow peas
80 g mesclun (or rocket, etc.)
2 tomatoes
2 teaspoons sugarbag honey
50 mL olive oil
4 g sea salt
2 g native pepper (or pepper)
2 finger limes (or 1 lime)
4 native pepper bread rolls (or simple rolls)

Method

- Boil kipfler potatoes. Angle slice.
- Blanch snow peas.
- Dice tomatoes.
- Slice 3 shallots
- Make dressing: Warm chopped shallots with sugarbag honey, add finger limes, season with salt and add olive oil.
- Season salmon with native pepper.
- Sear salmon.
- Crust kipflers in olive oil.
- Mix salad with dressing.
- Slice and toast native pepper bread rolls.

Serving suggestion

- Arrange salad in centre of plate.
- Slice salmon and arrange around salad.
- Drizzle with dressing.

IMAGE **GLENBO CRAIG**

EXQUISITE!

FEATURE
Propolis from stingless bee hives

Propolis is a mixture of resin and wax. Resin is collected from plants, but wax is produced by bees themselves. Stingless bees mix these two materials to form propolis (sometimes called cerumen).

Honey bees also secrete wax to use as a building material, collect plant resins, and mix the two to form propolis, but their nest is mainly constructed of pure wax. Commercial beeswax is from honey bees and is pure wax. Propolis from honey bee hives is also collected commercially and sold for various purposes. In this section, I will teach you how to harvest, process and use the propolis of stingless bees.

How to harvest and prepare propolis for use

Propolis can be readily harvested from honey sections at the same time as extracting honey. Usually only small amounts of wax are produced, about 200 g per hive per year. But some hives show a consistent tendency to store more in the honey section and these are good sources of the material. I use an American hive tool to scrape it out. This tool has a blade like a blunt chisel or stout paint scraper and is ideal for the purpose.

IMAGE **KATINA HEARD**

Propolis being scraped from the hive using a hive tool.

Large quantities (approximately 400 g) of propolis have been stored in this honey section, especially along the two sides.

Several forms of propolis can be observed in the honey section. It may occur as thick deposits, which are probably stores of the material or perhaps are laid down to insulate the hives. Often a dense network of connectives can be seen, especially adjacent to the slots into the brood box. These structures provide transport routes for the bees to and from the honey section. Or they may be sites where bees ripen honey. Finally, the pots themselves may be very thick-walled. All of these structures are good sources of propolis.

When transferring colonies from natural sites to boxes, there is often a lot of nest material left over; this is also a good source of propolis. Often only the brood is transferred to the hive box, along with some unbroken honey and pollen pots. All the damaged nest contents can then be processed into clean propolis after removing the food contents. It is also possible to extract propolis from hives that have died, but be careful, as whatever caused the deaths of the bees may have left some kind of contamination. I do not recommend that you use the propolis of a hive that has become infested by natural enemies.

Propolis will come out of the hive in variable condition. Sometimes it will be clean and pure. If it is from honey pots, it will be mixed with honey that needs to be washed off. Similarly, if the propolis formed the walls of pollen pots, it will be mixed with lots of that substance. It may be studded with dead bees. It may also be in irregular shapes.

If you wish to process your propolis into solid consistent lumps ready for use or storage, here's what you need to do.

First, wash the propolis in a large container of water such as a tub or bucket. Honey is easily washed off, but pollen takes a little more work. The pollen is held in pots and each pot will need to be worked with the fingers, under water, to disperse it. The propolis usually sinks and so the water can easily be drained off. Dead bees will float off in this process. You may need 3 or 4 changes of water before the propolis is clean. Then drain and dry it in a colander.

When you have accumulated a quantity of washed propolis, you can melt it to homogenise and compress it into a solid uniform lump. Melt the propolis using a simple solar melter consisting of a wooden box with a glass top. Place a metal pot containing the crude propolis in the melter. I do not recommend melting the propolis on a stove as the wax can ignite and cause a fire. The solar unit melts the propolis within a few hours on a hot day in full sun.

Now you can either produce propolis blocks or separate it into pure wax and remaining material.

To make propolis blocks, take the pot of propolis out of the melter and into the shade to cool. While it is cooling, stir it to mix the components, which will have separated. While stirring, break up any remaining large pieces of hard propolis. When the mixture is semi-cool, pour the propolis from the pot into a bucket of water to complete the cooling. Remove the lumps of cooling propolis from the bucket of water and work it in your hands to form the shape you require. The propolis can be very sticky, so wet your hands regularly to prevent it sticking and make it easier to handle. I form lengths, which I then chop into small blocks of about 25 g. These are a convenient size for most purposes. If you let it cool in a large lump, it will be difficult to divide into smaller bits when you need them.

You can also produce pure wax. This is a good idea if the propolis is very dirty. Leave the propolis in the melter until the wax has separated and is floating on the top. While the mixture is still hot, pour the wax off from the top into another container. The remaining material can be left in your bee yard to be scavenged. A common impurity in propolis from Queensland is the seeds of cadaghi, a eucalypt called *Corymbia torelliana*. These seeds are hard to remove and are a natural part of propolis, so I do not try to remove them.

Ways to use propolis

CANDLES: Wax that is separated from the resin can be used to make candles. John Klumpp has poured a native beeswax candle, which burns with a pleasant aromatic smell and sparkles every now and then, presumably as traces of resin enter the flame. D'Orbigny in Bolivia in 1839 made a similar observation: "When it is burned it emits an aromatic odor that is rather pungent and very agreeable. It is reserved up to the present for church uses."

Ointment made from stingless bee wax. IMAGE **GLENBO CRAIG**

OINTMENT:

Both Meg Davenport and Nicola Fuller use stingless bee wax to make ointment. Nicola mixes it with macadamia oil to make an extremely pleasant product for soothing dry skin and lips.

Cutting processed propolis into blocks for convenient handling. IMAGE **KATINA HEARD**

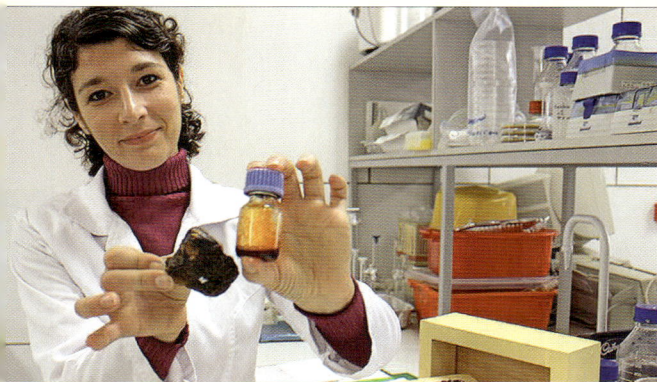

A

B

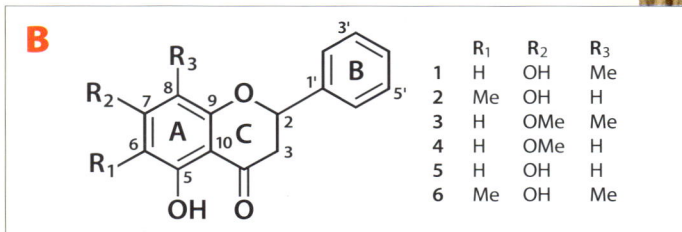

	R₁	R₂	R₃
1	H	OH	Me
2	Me	OH	H
3	H	OMe	Me
4	H	OMe	H
5	H	OH	H
6	Me	OH	Me

A A gumnut of the cadaghi tree cut open to reveal the yellow fruit resins (*ARROW*). IMAGE **AMANDA NORTON**

B The chemical structure of six flavanone compounds found in this resin.

MEDICAL USE: Flavia Massaro (above) and colleagues spent several years investigating the propolis of *Tetragonula carbonaria* and their favoured resin from the gumnuts of cadaghi trees. Flavia revealed much about its chemical composition and its bioactivity. She showed that this complex, enigmatic material may have therapeutic potential, for it relaxes coronary arteries and inhibits bacteria growth. Flavia analysed four resin deposits found in stingless bee hives and revealed the huge chemical differences in propolis from different plant sources.

BEEKEEPING AID: Propolis is a useful material for beekeeping. Use it to seal holes or to give a hive a valuable resource if it is struggling. Propolis is also useful for making partial closures for entrances following a hive division. Read how to do this in "Protecting your hives from natural enemies" in Chapter 14.

CRAFT MATERIAL: Propolis can be used for Aboriginal hand crafts. A major potential use for propolis is to form the mouthpiece of the **didgeridoo**, a traditional musical instrument. Until recently, propolis was very difficult to obtain and honey bee wax was used as a substitute. However, stingless bee propolis is the authentic and superior product. It has a great smell that enhances the experience of playing. Demand is expected to grow rapidly as awareness increases. Propolis is a rare product and should be valued as such. I currently sell it for about $250 per kg wholesale, or $10 for a 25 g block. A block of this size is enough for one or two didgeridoo mouthpieces.

How to make a didgeridoo mouthpiece

Take about 25 g of propolis. Soften it with heat (drop it in warm water or put it in the sun). You should be able to work it easily into the shape you desire. I form it into a sausage and then join the ends to make a ring the same diameter as the mouthpiece. Push it on the end of the didgeridoo and push it hard down to form a good bond. Smooth the inside and outside edges so that it sits tightly on the didgeridoo and makes a good seal for the player's mouth. It will harden with age, but it never really sets and can melt on a very hot day.

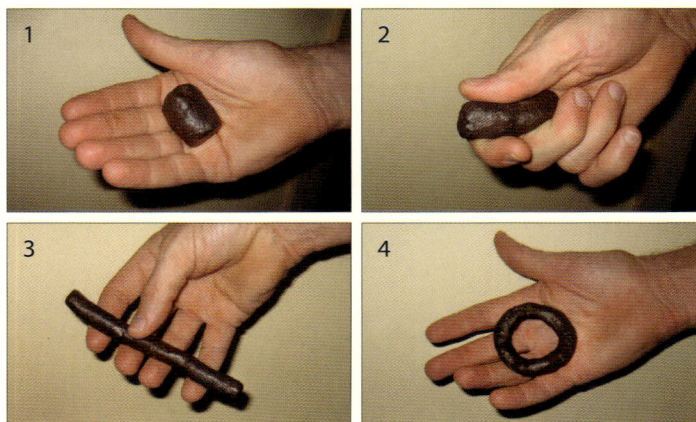

Make a didgeridoo mouthpiece by working the propolis into a ring and pressing it onto the end of your didj.

13 Managing and protecting stingless bee hives

Stingless bees collecting resin from a fruit (gum nut) of the eucalypt, *Corymbia torelliana* (cadaghi). Is this tree a friend or a foe of stingless bees?
IMAGE **LAURENCE SANDERS**

Hives can be fed with sugar solution, but this is rarely necessary.

If you plan to keep colonies **outside their native range**, you must introduce multiple colonies to ensure genetic diversity. Within the native range, one colony is sufficient because males from other colonies will find and mate with your queen.

Breeding stingless bees for **bee improvement** has not started in Australia. We need to decide what we want of our bees, then find colonies with those traits and, finally, develop techniques to introduce those traits into other colonies.

You can determine the **health and strength** of a hive by its foraging activity, weight, smell, sound, and size of brood and food storage.

Stingless bees can bite! But there are ways to protect yourself if it bothers you.

Stingless bees **attack each other's colonies** to gain access and to install their own queen. This causes fighting swarms that are hard to manage but will not result in your hive dying.

Stingless bees suffer in **extremely hot or cold weather** and need to be protected from temperature extremes.

Stingless bees **collect resin and seeds** from the fruits of the **cadaghi tree**, and help to disperse cadaghi seeds. Although some beekeepers have blamed cadaghi resin for hive deaths, evidence indicates it is harmless to hives.

The flowers of the **African tulip tree** are poisonous to bees and other insects, but few bees get caught in these blooms.

In Chapter 8 we covered some of the more basic topics you need to know about to start keeping stingless bee hives. In this chapter, we explore more advanced topics. You will learn how to keep your hives healthy, and how to protect them from a range of threats that they may face.

Chapter 14 then goes on to give more detailed advice about how to protect your stingless bees from the threats posed by natural enemies.

13.1 Feeding bees

A healthy colony of native stingless bees can last without food for a long time. I weighed two colonies in a glasshouse with no forage and they lost about 14 g of weight per day. This suggests that a weak hive with only about 1 kg of food should have enough stores for 74 days. A strong colony with 4 kg of stored food should be fine for maybe three or four times that long. I cannot base this on detailed data (hence my use of the word "maybe"), but it gels with my experience providing hives for an enclosure at Melbourne Museum. I send them strong colonies, which feed only a little in the enclosure, and they last about 9 months before their weight gets critically low.

So stingless bee colonies can survive long periods without food. However, in some situations, it may be helpful or necessary to feed your stingless bees. They need both carbohydrate (sugar) and protein. It is easier to feed them a carbohydrate source than a protein source, and feeding them carbohydrate alone is still beneficial, because it allows foraging bees to concentrate on collecting pollen.

Stingless bees also need a resin supply.

13.1.1 SUGAR

A sugar solution provides a good carbohydrate source.

Make up a 50% solution by dissolving 100 g of white sugar in 100 ml of water. This solution can be provided either at open feeders placed in the field for the bees to find and exploit, or in feeders placed directly at the hive entrance. Honey bee feeders also work for stingless bees (Figure 13-1).

Anne Dollin and John Klumpp provide some designs for hive feeders in their publications. For a very simple field feeder, use a yellow or blue sponge (these colours are most attractive to the bees, not red). Simply dip the sponge in the sugar solution and place it on a saucer. This food may be found by honey bees and other animals, so try covering it with a mesh that has an aperture size of about 3 mm, which will allow stingless bees to enter, but not larger animals. The sugar solution will be found more quickly by the bees

FIGURE 13-1 Stingless bees at a sugar-water feeder. IMAGE **JULIA GROENING**

if you provide a scent such as vanilla, lemon, rosewater, etc. You can add the scent to the solution, but be very careful to add only a tiny amount, or it could poison your bees. It is safer to provide the scent separately on an absorbent material.

It may be tempting to feed bees honey rather than sugar water. But sugar water may be preferable: it avoids the risk of transferring diseases, and research suggests that it is preferred by bees, and may benefit them more than honey.

In laboratory trials, Giorgio Venturieri and I gave *T. carbonaria* worker bees the choice of a sucrose solution (made from cane sugar) or honey (from stingless bees), and found that they preferred the sucrose. We then fed them without choice on either a sucrose solution or honey and found that they lived longer when feeding on the sucrose. We also gave them a choice of 34%, 54% or 74% sucrose, and they preferred the lowest concentration. We compared our results with some studies on honey bees and found that honey bees also prefer sucrose solution to honey.

This result is counter-intuitive; you would expect that bees' own stored food would appeal to them more than a less natural alternative. We looked at possible reasons for this and found that, with age, honey undergoes biochemical changes that produce toxic compounds (e.g. the level of hydroxymethylfurfural increases), which may deter bees. (Read more about this in "Honey composition" in Chapter 12.)

13.1.2 PROTEIN

Feeding bees a protein source is more difficult.

You can feed stingless bees with pollen from honey bees, which is readily available in health food stores. Grind the pellets and place the resulting powder in a container near the hive. Mix in a little sugar solution to help attract bees.

You could also try one of the artificial protein sources available for honey bees and bumble bees.

Giorgio Venturieri has developed the use of fermented soy milk paste as a protein source. Make a paste by mixing soy milk powder with water and a little sugar. It seems to only be acceptable to stingless bees if it is fermented. Take a few pollen pots from a stingless bee hive and mix the pollen into the paste. This hive pollen will inoculate the mixture with microbes and initiate fermentation. Allow it to ferment for about one week depending on the ambient temperature. Feed this to the bees either internally or externally to the hive. The best way to feed it is to make balls, coat them in propolis collected from a hive, and introduce those into the hive. It appears to be palatable and beneficial to bees.

I recommend feeding your bees pollen or an artificial protein substitute only when there is no alternative natural flower food source available, as naturally collected pollen is likely to be a superior protein source.

13.1.3 RESIN

In addition to pollen and nectar, plants provide resinous material for stingless bees. Resin has multiple benefits in the nest, including use as a building material, protection against pathogens, social defence against predators, and as a nest entrance repellent.

Sara Leonhardt and colleagues have shown that a diversity of resins is beneficial. For example, some natural plant resins are particularly repellent against hive beetles, while others are more effective against certain pathogens. Therefore, in addition to needing diverse floral resources, bees also need diverse resin sources. These sources are usually naturally available in bushland or urban environments, but could be limiting on farms or areas of low plant diversity.

If you can, make a list of local plant species that produce resins that are attractive to bees and plant those species on your property. It is also possible to obtain propolis from keepers of stingless bees and provide that to your bees.

13.2 Maintaining genetically viable populations

Bees are considered particularly vulnerable to genetic impoverishment due to their haplodiploidy. (Read more about this in "How is gender determined in bees?" in Chapter 3.)

When bee populations lose genetic diversity, chances increase that queens will mate with males that share identical sex alleles. This results in strange creatures called "diploid male bees", instead of the usual haploid males. (Read more about this in "Diploid males", also in Chapter 3.) Theoretically, production of diploid males is undesirable, because a large proportion of the young bees destined to become female workers will end up developing into males, which do not work or contribute in any way to the colony. This will slow down colony growth and increase colony mortality.

At least, this is the prediction. In a recent study from Brazil, researchers artificially imposed a genetic bottleneck by setting up a population starting from only two founder colonies of stingless bees, and continued breeding from them for a period of over 10 years in a location outside the bees' natural area of occurrence. The study revealed a reduction in genetic variation and an increase in sterile diploid males. But these genetically impoverished populations continued to successfully breed for the 10 years of the study. This indicates that, in stingless bees, breeding from a small stock of colonies may have less severe consequences than previously suspected.

What are the consequences of this? First, it is a potential problem only if you move your colonies outside their native range to a location where they can't naturally find another colony of the same species. This is not recommended anyway. But, if you do it, then ensure that at least two genetically dissimilar colonies are taken to the new site.

13.3 Breeding better bees

Animal breeding is the art and science of changing the traits of animals from generation to generation in order to produce desired characteristics. It may be called "animal improvement" to distinguish it from the breeding of animals to increase their numbers.

Honey bees are constantly being improved in breeding programs in which queens and drones that have specific beneficial characteristics are mated. Honey bees of different races are also hybridised to produce desirable qualities. The desired traits include resistance to pests and diseases, good honey production, prolific breeding, low tendency to swarming, and good temper. Some honey bee queens are bred using artificial insemination, a technique that gives complete control over mating and is an important tool for rapidly improving honey bee characteristics, maintaining those improved characteristics, and producing breeder queens.

Beneficial characteristics can also be selected and maintained by flooding the immediate area around mating apiaries with honey bee drones of known beneficial characteristics. Commercial keepers of honey bees routinely re-queen every year or two to keep hives in top production mode led by a young, healthy queen with a better egg-laying capacity than an older queen.

I expect that, in the future, efforts will also be made to improve stingless bees. However, before any breeding program takes place, we need to clearly define its goals.

One obviously desirable trait is greater productivity. In fact, we are already implicitly "breeding" for this in the propagation of stingless bee hives. More productive hives are divided more often, so their superior genes become more common. However, a more focused breeding program could yield benefits if a specific desired trait is recognised, for example, for colonies with a reduced tendency to engage in fighting swarms. So the first steps are to decide what we want of our bees and to then identify and isolate colonies with those traits. We could then develop techniques to introduce those traits into other colonies.

13.4 How to determine the health and strength of a stingless bee hive

You can use a number of factors to determine whether a stingless bee hive is healthy, including foraging activity, weight, size of brood and food storage, smell, sound and garbage removal.

FORAGING ACTIVITY. Assess foraging activity by counting the number of foraging bees in a specific period (e.g. one minute). You will need an audible timer such as one in your watch or phone so that you know when one minute has passed without having to take your eye off the hive entrance. Count each bee as it enters or leaves the hive, whichever you find easiest.

Choose a time when the temperature is suitable (between 22°C and 35°C). Light levels also affect foraging, so try to do your count on a sunny day. Also try to avoid a very windy day. For *Tetragonula carbonaria* and *T. hockingsi*, you should be able to observe at least 20 incoming bees in the one-minute period. At a strong hive entrance, you will typically observe between 30 and 60 bees per minute returning. You may observe up to 100 incoming bees at a very strong hive at optimum temperature, under a clear sky and with good forage available. If you use a five-minute period, the estimate will be more accurate but, of course, you will have to multiply by five the target figures outlined above.

This technique estimates the colony strength of foragers, which correlates closely with other measures of colony strength such as brood volume. But it is not a perfect correlation, as a healthy hive could have a small foraging population and a hive with many foragers could have some other problem. Also, this technique works well for the species of *Tetragonula* but the activity of *Austroplebeia* colonies is lower and more variable. (Read more about this in "Foraging strategies" in Chapter 5.) If conditions are good and there is an attractive food source nearby, then you may see 15–20 foragers per minute from a colony of *Austroplebeia*.

WEIGHT. The weight of a hive is a good proxy for its health. Of course, you need to know the weight of the empty box and subtract this from the overall weight. A net weight of 2–4 kg indicates a healthy hive.

SIZE OF BROOD AND FOOD STORAGE. It is relatively easy to open a hive of honey bees to observe the brood area, but the architecture of a stingless bee nest makes the concept of brood area meaningless. Instead, we need to estimate the brood volume. You may be able to estimate the volume by measuring the brood diameter through a window, or by opening the hive. The brood diameter of a strong hive of *T. carbonaria* will be greater than 100 mm, while that of *T. hockingsi* will be 120 mm or more. Also note the volume of the hive's food reserves.

SMELL. A stingless bee hive should smell resinous (like eucalyptus oil) and not pungent (like rotten fruit). A bad or off smell indicates poor health or possible attack by natural enemies.

SOUND. A buzzing sound at the hive entrance also indicates good hive health, although you will need to put your ear close to hear it. This is a useful technique to assess a hive at night.

GARBAGE REMOVAL. Garbage pellets being removed from a hive in the mandibles of workers is a positive sign because it indicates that the colony is actively producing adults.

13.5 Protecting yourself against the defences of stingless bees

Stingless bees have minimal ability to defend themselves against humans and other large predators. Even blocking an entrance with a finger will not incite any defensive reaction. But, when a nest is opened, they may defend themselves vigorously by flying into the air, identifying the source of aggression, landing on the attacker, and crawling onto softer parts of the body, such as eyelids (Figure 13-2), underarms, crotch, and

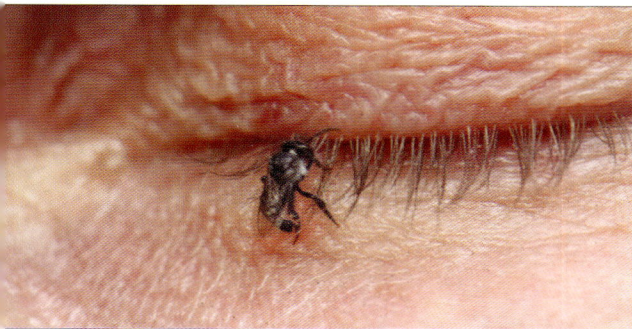

FIGURE 13-2 A stingless bee biting the beekeeper's eyelid. IMAGE **GLENBO CRAIG**

inside elbows. Stingless bees can bite with their mandibles and cause discomfort or even minor pain. The repeated biting of many bees can leave small, red skin lesions that may remain red for several hours (Figure 13-3).

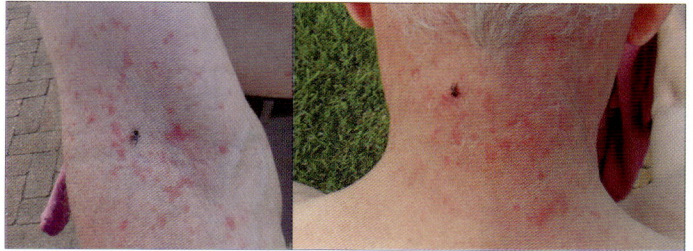

FIGURE 13-3
Bites by stingless bees leave red marks on the skin. Here, defenders of a *Tetragonula hockingsi* colony have bitten the beekeeper's inner elbow and neck. IMAGE **DANIELLE LELAGADEC**

You can adopt various protective strategies, such as insecticide, suitable clothing, and working at the right time of day. Smoke does not work. Strategies to protect yourself are discussed in more detail below.

13.5.1 SMOKE?!

Smoke is routinely used to pacify honey bees before opening their hives. But keepers of stingless bees know that it has no effect on their bees. Honey bees respond to smoke because it stimulates them to prepare to abandon their nest site to escape fire. In this state, they are not motivated to defend their nest and are easier to handle. But stingless bees will not abscond from a nest site (because the reigning queen cannot fly once she has mated, and the colony will not abandon its queen), and so they have not evolved the behaviour of dropping defences in the presence of smoke.

13.5.2 CLOTHING

Dark clothing seems to stimulate a strong defensive response from stingless bees, so avoid wearing black when opening hives. If you have dark hair, the bees may attack and entangle themselves in it. Try to cover your hair with a light-coloured hat. Some keepers wear a veil to protect their face; this can be useful if you expect a strong defensive reaction. I prefer to wear long trousers when working with stingless bees. If you are wearing shorts, I have just a few words of advice: "tight-fitting underpants"!

13.5.3 TIME OF DAY

Some keepers of stingless bees prefer to work at night when the bees are less likely to fly and bite. But I find that, at night, bees will often leave the hive in large numbers and walk randomly, making them susceptible to being crushed underfoot.

13.6 Fighting swarms

Stingless bees of the *Tetragonula* genus often display an extraordinary and spectacular behaviour, in which a cloud of flying bees swirls in front of a nest entrance (Figure 13-4). The cloud comprises invading bees mixed with defending bees from the nest.

The swarm may change so that many bees orientate with their heads towards the hive entrance (Figure 13-5). Or many bees may aggregate on the outside of the hive and stand tall while fanning their wings, appearing to be emitting pheromones (Figure 13-6).

Sometimes the swarming ends when the defending bees temporarily seal the entrance of their nest with propolis (Figure 13-7). This may be the end of the conflict, and a normal state will return within a few days. On the other hand, swarming may escalate into a full-blown bee battle. In this case, the swarming bees grip each other with their mandibles and fall to the ground locked together. These warriors will never release but will die in mortal combat (Figure 13-8). Thousands of bees may perish in this way. The ground beneath the entrance may be covered in a carpet of bees a centimetre thick; 7,000 corpses littered one battleground (Figure 13-9). All the fighters are female workers. Some male bees may also be present, but only flying; apparently, they are lovers, not fighters!

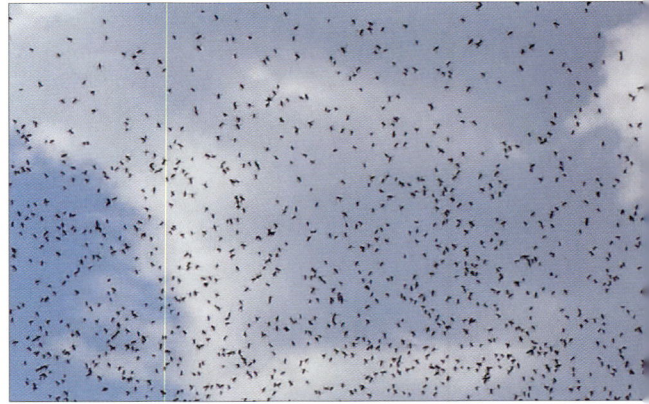

FIGURE 13-4 A swarm of bees circulating in front of a hive entrance. IMAGE **RICHARD TANNER**

FIGURE 13-5 Swarming bees orientated towards the entrance of a hive. IMAGE **TOBIAS SMITH**

FIGURE 13-6 Bees crawling on the hive, some appearing to emit pheromones, a common behaviour during fighting swarms.

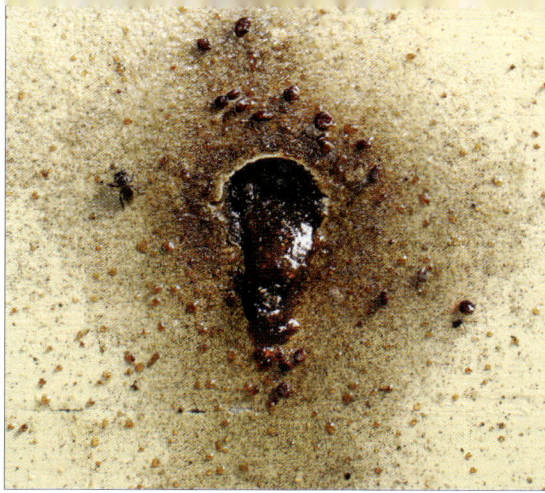

FIGURE 13-7 A hive entrance blocked with propolis during a fighting swarm.

Sometimes, towards the end of a fighting swarm, immature bees and young callow adults are dragged from the defending hive. The swarms may continue for a few weeks. Sometimes they cease and then start again weeks or months later. They can occur at any time of the year, but mostly in summer. Normal foraging activity usually ceases during fighting swarms, but colonies do not normally die out as a result of fighting.

FIGURE 13-8 Bees locked in combat beneath a hive entrance.

IMAGE **PAUL CUNNINGHAM**

FIGURE 13-9 Mass of dead and dying fighting bees, 3 cm deep, on the ground beneath a hive entrance. IMAGES **JEFF WILLMER**

Of all the strange behaviours of stingless bees, swarming has to be one of the most complex, poorly understood and variable, not to mention traumatising for those witnessing their bees committing murder. What is going on? Is there anything we can do about it?

Fighting swarms involve bees from a defending hive (where the battle takes place) and an attacking colony (the location of which is usually unknown). The attacking colony is attempting to take over the defending one by installing its own queen. How do we know this? The work of Ros Gloag and colleagues provides strong evidence. After death, pairs of fighting workers were separated and genetically fingerprinted. Two genotypes were detected within each fighting pair. One of the genotypes always matched the bees in the defending nest. The other genotype was that of another colony, the attacking colony. Sometimes multiple other colonies joined the attack but, generally, fights involved one defending colony and another attacking colony. (Hence some beekeepers call swarming a "home invasion".)

Ros' work was done with the species *Tetragonula carbonaria*: one colony of *T. carbonaria* attacking another colony of the same species (*intra*-species fights). But more recent work by Paul Cunningham and colleagues shows that swarming can also happen between species, creating *inter*-species fights. The two species involved were *T. carbonaria* and *T. hockingsi*, closely related species. As in Ros' experiments, fighting bees were genetically fingerprinted, revealing that they were from one defending and one attacking colony. However, after a few months, the genotype of the defending colony had changed to that of the attackers, proving that usurpation had occurred.

A final experiment showed that this happens quite frequently. Over a five-year period, I worked with colleagues to monitor annually the identity of 250 colonies. During this period, species changes took place 46 times. These were mostly cases of *T. hockingsi* taking over *T. carbonaria* nests. The larger hockingsi appear to be better usurpers than carbonaria (Figure 13-10) but, on two occasions, it happened the other way.

Male bees are also seen close by when a fighting swarm is taking place, but they do not participate in the fighting. Rather, they may loiter close by in

FIGURE 13-10 A larger *Tetragonula hockingsi* (RIGHT) biting a smaller *Tetragonula carbonaria* (LEFT).
IMAGE **GIORGIO VENTURIERI**

the hope that a mating opportunity is imminent. This is because, after a colony is successfully taken over, the attacking colony will introduce its own queen and she will need to take a mating flight.

Colony takeovers are known from other social insects but are not common and the scale of devastation is generally not as great as among stingless bees. Usually, animals avoid heavy losses by tests of strength. But, in our native *Tetragonula* stingless bees, the battles are monumental in scale and brutally fierce. Stingless bees may be vegetarians and lack stings, but they show extremely aggressive behaviour when it comes to occupying the nests of other colonies.

13.6.1 WHAT CAN WE DO ABOUT FIGHTING SWARMS?

When a fighting swarm reaches a crescendo with many bees fighting, we can use the opportunity to capture the attacking bees into a trap hive, at the same time taking the pressure off the defending hive. Learn how to do this in section 10.5 on Trap hives. It doesn't always work, but is definitely worth trying.

13.6.2 DRIFT FIGHTING

Research has shown that fighting swarms can also be triggered by swapping the positions of two hives during a period of active foraging. This causes the bees from one hive to attempt to enter the other. The defending colony immediately launches an aerial defence. The foraging bees are not really attacking the hive, just trying to go home. But they are perceived by the defenders as attackers and engaged in combat. Within a day, the non-nestmates are dead. The defenders will stay on high alert for several days. During this period, defending bees will attack each other but stop once they recognise that their opponents are actually nestmates. This is called **drift fighting**.

The research helps us to understand fighting swarms and to manage them. For example, we observe these swarms when hives are taken onto a farm for pollination. In the new location, foragers may get lost and **drift** into the wrong hive. In honey bees, drift does not cause too many problems, because non-nestmates are generally accepted. Honey bees do not try to usurp each other's hives, so they are not defensive. But stingless bees are very sensitive to non-nestmates entering their hives, because it indicates an attempt at usurpation.

You can manage your hives to minimise drift. Space out hives and mark them so that they look different. Use different colours and symbols on the front, top or roof of each hive. Bees will learn the pattern on their hive and be less likely to return to the wrong nest. Also place the hives so that natural landmarks guide the bees back to the correct hive. Do not place a row of hives along a homogenous fence line.

13.7 Protecting colonies from weather extremes

All social bees have strategies to maintain favourable nest temperatures. (Read more about this in "Temperature regulation in social bee nests" in Chapter 4.) However, stingless bee hives are less resistant to extreme heat and cold than honey bees. Therefore, stingless bees need extra protection from the elements. In this section, I cover the two main climatic threats to hives, heat and cold, and give you some ideas on how to deal with them.

Heavy rain is not usually a problem for a well-made hive. Flooding is a risk, but I have seen stingless bee hives survive complete inundation and being filled with water. If this happens to one of your hives, simply empty the water out through the main entrance or drill a hole in the bottom of the hive to allow it to escape.

13.7.1 EXTREME HEAT

Stingless bees are not capable of the evaporative cooling displayed by honey bees. This has two consequences for beekeeping. First, water needs to be provided to honey bees in hot weather, but not to stingless bees. Second, stingless bee hives need to be well protected from heat by insulating

hive boxes and placing them in cool positions. With good insulation and placement in the shade, the nest will be protected from ambient extremes of heat (Figure 4-26 B).

Insulation is all about preventing peak heat from penetrating the box. In some recent extreme heat events (such as in parts of Queensland on 4 January 2014 and in Sydney in 2013 and 2014, when temperatures reached 45°C), many colonies reportedly died. Giorgio Venturieri showed that workers of *T. carbonaria*, in laboratory conditions, died after 10 minutes when the temperature was over 39°C (Figure 13-11). *Tetragonula hockingsi*, which naturally occur in more tropical areas, were more resistant to heat. The colonies that died during the heatwaves probably reached this lethal temperature in the brood chamber. Hives die instantly when both the adult and brood populations are killed (Figure 13-12).

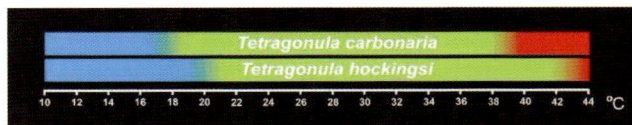

FIGURE 13-11 Temperature tolerances of two common species of stingless bees. Note that *Tetragonula hockingsi* is better adapted to higher temperatures than *Tetragonula carbonaria*. **BLUE:** bees stop moving but do not die, even after 48 hours. **GREEN:** bees fly. **RED:** bees die.
ARTWORK **GIORGIO VENTURIERI**

FIGURE 13-12 Two hives instantly killed by heat on a very hot day. On the left, an opened hive of *Tetragonula hockingsi*; on the right, *T. carbonaria*. These previously very strong and healthy hives were positioned in the sun.

FIGURE 13-13 Lines of bees standing around the entrance and fanning their wings is a sign of overheating. IMAGE **CAROLINE GARDAM**

FIGURE 13-14 A hive well positioned in the shade of a tree and with an overhanging roof.

A colony can also show a prolonged decline in adult bee numbers, probably because adults survived but the brood died so no new adults are coming through. Sometimes we see an instant loss of adult activity but then the hive appears to revive, indicating that the brood survived, and adults continue to emerge. But, sadly, these hives also usually die eventually.

An obvious indicator of overheating is the appearance of adults radiating in lines facing away from the entrance and fanning their wings (Figure 13-13). They are making a desperate attempt to draw cool air into the hive. You need to take immediate action if you observe this.

The following strategies will help your hives to deal with heat.

● Ensure hive boxes provide good insulation. (Read more about this in Chapter 9, "Constructing hive boxes for stingless bees".)

● Ensure the hive has a ventilation hole in addition to the main entrance.

● Position the hive in a shaded position, especially in the afternoon. Direct sun on the hive can be dangerous in summer. Shade may be provided by trees, walls, etc. (Figure 13-14). Metal roofs are proven to cool the walls on hives in the sun (Figure 13-15). (Read more about this in "Metal roofs" in Chapter 9.)

● In emergency situations, when extreme events are predicted, take action:

– Drape a wet towel over the hive. Immerse one or both ends of the towel in a bucket of water (Figure 13-16). The evaporative effect of the water may help to cool the hive. Keep the entrances clear.

– Move the hive into an air-conditioned room. If extreme temperatures are forecast, you could close the hive the night before and take it into an air-conditioned room. In extreme heat conditions, save the colony by moving them during the day. This is a desperate measure because you will leave some workers behind, but the foraging bees are only a small proportion of the colony's population and will be replaced within days if the colony survives.

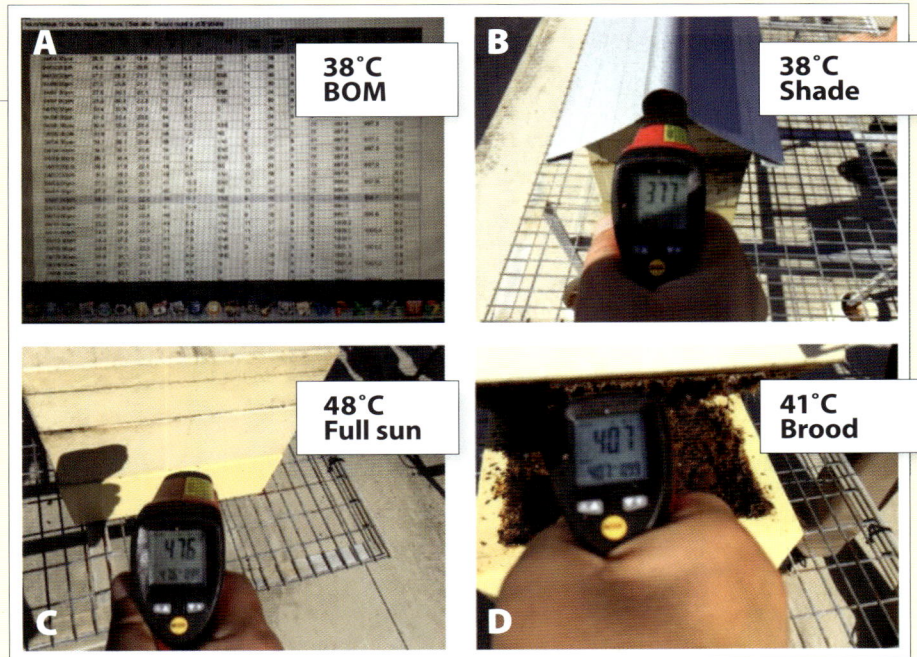

FIGURE 13-15

Temperature on a hot day:
A: Ambient (Bureau of Meteorology).
B: A hand laser thermometer measurement from the top of a hive under a roof.
C: The side of a hive in full sun (10°C more than in the shade).
D: The internal brood temperature (close to the lethal temperature).

IMAGES **GIORGIO VENTURIERI**

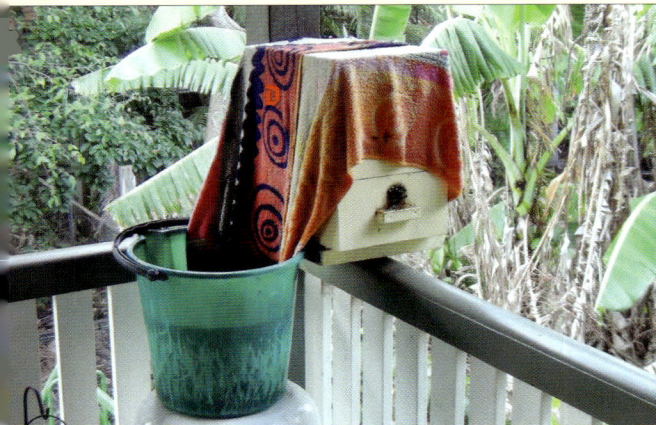

FIGURE 13-16 A hive with an emergency cooling system of a wet towel with one end in a bucket of water.

13.7.2 EXTREME COLD

Problems due to low temperature can manifest themselves in two ways. Each requires a different response.

- The bees die from low temperatures. This is likely in inland areas with extremely low night temperatures. An area that experiences extreme frosts will be risky for stingless bees. You can mitigate this problem with good hive box insulation. (Read more about this in Chapter 9, "Constructing hive boxes for stingless bees".) When extreme cold nights are predicted, move the hive into a heated room overnight. Artificial heating with thermostat control is used by Ian Coots, who has kept a *T. carbonaria* hive in Geelong, Victoria for five years.

- The colony runs out of food because it is too cold for bees to forage during the day. This is likely in southern areas that have low day temperatures for an extended period. You can help the bees by artificially feeding them. Also place the hive in direct sun to warm it and improve bee activity and brood growth.

The question of how low a temperature a hive can tolerate is relevant and important if you are keeping colonies in cooler areas. It sounds like a simple question, but it is surprisingly complex. First of all, you cannot consider cold temperatures on their own but must also consider how long the temperature stays low. For example, you can put bees in the refrigerator, take them out after a few hours and they will revive. But if you left them in there for days, you might get a different result. So cold is not just about temperature, but also about time; a mathematician would say that there is a relationship between bee survival and a "time temperature product". In other words, long periods of cold are going to have a greater effect than a short cold snap.

Giorgio Venturieri showed that, at 18°C, most individuals of *T. carbonaria* stop moving but remain standing (Figure 13-11). We have long known that they do not leave the nest below 18°C, but now we also know that they cannot even walk at this temperature. At 14°C, they cannot stand but fall onto their sides. *T. hockingsi* are even less tolerant of cold, displaying the same effects at about 2°C higher. So temperatures below 18°C inside the nest should be avoided. When this temperature is approached, minimal metabolic warming occurs, brood rearing and care ceases, and the nest will not be guarded.

These are the responses of bees when the temperature reaches a certain point *inside* the hive. But, of course, the temperature there differs from that on the outside. Here are some reasons why:

● Insulation of the box. If a box is a better insulator, then it will remove the peaks and troughs of temperature variation (Figure 4-26). The factors that will affect how well a box is insulated include the type of material used, the thickness of the material, and any points where the insulation is compromised (entrances, gaps).

● Insolation (amount of sunlight hitting the box). Peter Clarke reminds us of the importance of a sunny position for hives of *T. carbonaria* in Sydney, which is getting towards the cooler end of the natural range for this species. Peter has found that it is important for survival through winter that the hives receive as much sunlight as you can give them.

● Air movement. If the hive is in a windy position, air currents may strip away colony warmth more quickly.

● Size of the nest population. A large colony will generate more metabolic heat, which will help to keep the bees cosy inside. This metabolic heat then interacts with nest insulation that keeps the heat in. Insulation is also improved by more food stores around the brood and propolis deposits around the nest.

● Temperature variation in the nest. Bees may be able to migrate to warmer parts of the nest when it is colder. I have seen the honey section of a hive deserted on cold mornings as the bees move into the brood area.

So there is no simple answer to the question of how low a temperature a colony can survive. But an understanding of the various contributing factors and a willingness to take action will help to keep your hive alive through winter. In addition to the heating and feeding remedies mentioned above, you may need to send your hive somewhere warmer.

13.7.3 RE-USING A DEAD HIVE WITH ITS RESOURCES

If your hive does die due to weather extremes, and you are confident that it has not yet been infested by natural enemies, you can use it to divide another live hive. (You can also do this if a hive dies due to a failure to re-queen or for an unknown reason.) I have done this successfully many times. The two halves usually grow very strongly as a result of the availability of food stores and structures such as the entrance tube, and are ready to divide again soon. Follow this procedure:

1 Check for pests, such as the larvae of hive syrphid flies, hive phorid flies, small hive beetles, ants, brood disease, etc. If some pests are present do not proceed but clean out the box and start again from empty.

2 As a precaution, put the hive into a freezer for a few days to kill any undetected pests. After removing the hive from the freezer, allow it to return to ambient temperature, then proceed.

3 Remove the dead brood, but leave the stored food in place.

4 Couple each half to the two halves of a hive that has a fully developed brood. You have saved the food reserves of the dead hive and recouped some of your losses.

An alternative use of a dead hive is to attract a fighting swarm. (Read more about this in "Trap hives" in Chapter 10.) Note that, with either of these methods, you run the risk of transmitting disease. Although this seems to be a minor risk with stingless bees, you must allow for the possibility.

13.8 Protecting colonies from insecticides

Bees are susceptible to insecticides. Pest sprays used inside your house should not harm your bees because the sprays are unlikely to enter the hive. But treatments for termites or other pests that involve heavy applications of persistent insecticides around the house are a threat. If you are having one of these treatments, move your hive away for a few weeks.

Garden sprays may also be dangerous for your bees if one of their flowering forage plants is sprayed. If you suspect this is the case, move the hive away for a few days after spraying, or close it and move it to a cool spot. Also consider alternative pest control options such as pest traps for fruit fly, etc. These traps will not usually pose a threat to bees. Even if a trap uses an insecticide bait, they will not be attractive to bees and so pose no risk.

For strategies to deal with insecticides on farms, see "How to minimise the effects of pesticides" in Chapter 16.

13.9 Cadaghi resin: threat or resource?

FIGURE 13-17 Cadaghi flowers are much loved by stingless bees, but it is a couple of months later, when the resulting fruits mature, that the really interesting part begins.

Keepers of stingless bees in Australia are reminded daily of the intimate relationship between their bees and a eucalypt tree called cadaghi or *Corymbia torelliana*. Around October of each year, the bees actively forage on the flowers of the cadaghi, which provides a useful source of pollen (Figure 13-17). Seedlings of this tree are often seen germinating below hive entrances (Figure 13-18). The seeds of the cadaghi (red objects about 2.5 mm in diameter) are also often seen around the hive entrance, sometimes in massive numbers (Figure 13-19). And when you open a hive, you will often see lots of cadaghi seeds, usually combined with a resin that resembles mozzarella cheese in colour and stringiness (Figure 13-22, 4-21).

What on earth is going on?

FIGURE 13-18
Seedlings of the cadaghi tree below the entrance of a *Tetragonula carbonaria* hive.

FIGURE 13-19
Seeds of the cadaghi tree around the entrance of a *Tetragonula carbonaria* hive.

So what is going on?

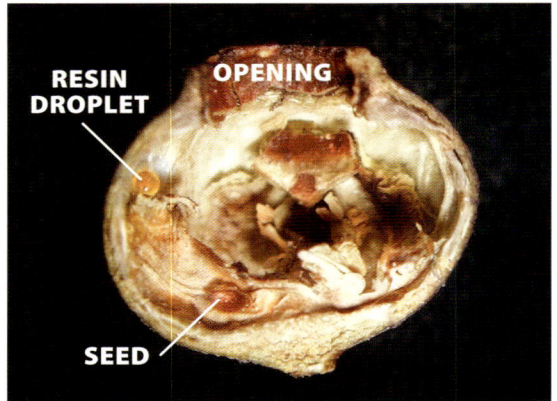

FIGURE 13-20 A cadaghi fruit (gumnut) dissected to show seeds and resin droplets inside.
IMAGE **HELEN WALLACE**

FIGURE 13-21 A *Tetragonula carbonaria* bee with a cadaghi seed stuck to its leg. IMAGE **DANIEL KLAER**

Careful work by Helen Wallace solved the mystery. Helen first identified the seeds as belonging to *Corymbia torelliana*. She then dissected the mature gumnuts (hard, hollow fruits) and discovered resin glands producing sticky droplets inside (Figure 13-20). She observed that stingless bees were attracted to the resin, entered the gumnuts to collect it and, in the process, often ended up with seeds stuck to their bodies. They then flew to their nests. But the weight of a seed (about half that of the bee itself) was such that they struggled to carry it (Figure 13-21). So, before, during and after the flight, they tried to remove the seeds, dispersing them as they went. A small proportion of bees still had a seed stuck to them when they reached the nest.

Clearly, a mutually beneficial relationship has evolved between the bees and the trees. The bees provide long-distance dispersal of seeds (at least 300 m and perhaps more than 1 km) and are rewarded with a viable source of nest-building resin. At least five species of stingless bees have been observed dispersing cadaghi seeds in the wild.

The season during which bees can collect cadaghi resin typically lasts two to three months, starting with fruiting in December and ending when the bees have removed the resin from the fruits. Generally, only a small proportion of foraging bees will return with resin, and only one-quarter of these will also be carrying a seed.

However, these figures are variable; occasionally, in the peak season of cadaghi fruiting, a larger proportion of the bees will be collecting resin. The bees stockpile the resin in the hive, later combining it with wax to make propolis, which they use to build structures inside their nests.

Some cadaghi seeds are incorporated into the nest structures (Figure 13-22) but many are removed and dumped outside. In the case of *Tetragonula carbonaria*, the seeds are deposited on the outside of the entrance and remain there. *T. hockingsi* bees move the seeds well away from the entrance and do not allow them to accumulate there. Even *Austroplebeia australis*, which does not use a lot of propolis in the nest, collects the resin of cadaghi, as evidenced by the seeds inside their nests (Figure 13-23).

Labels on image: RESIN, SEEDS, RESIN, NEST STRUCTURES, BROOD

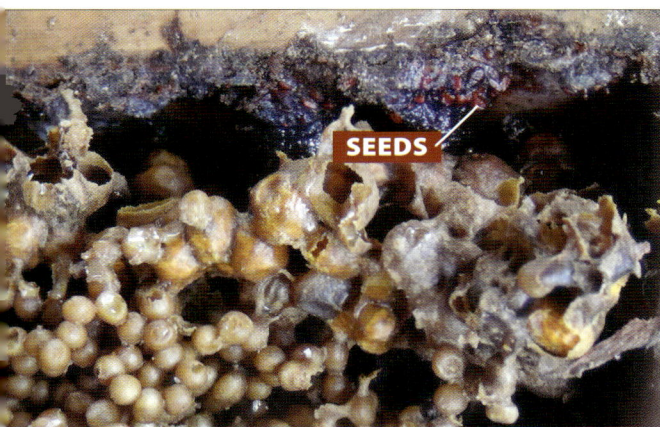

Labels on image: SEEDS, RESIN

FIGURE 13-23 Red cadaghi seeds in an *Austroplebeia australis* nest.

FIGURE 13-22 Seeds and resin of the cadaghi tree inside a *Tetragonula carbonaria* hive. The bees have used this material to build nest structures. IMAGE **GLENBO CRAIG**

The resin of cadaghi has been the subject of much discussion and angst. Although native to far north Queensland, this tree has been planted as far south as Newcastle in New South Wales. It is abundant in many areas where stingless bees are kept, and local bee species take advantage of its now widespread distribution. Cadaghi resin can accumulate in large deposits in the hive. Many keepers of stingless bees warn about the dangers that this resin poses to hive health. Some claim that the resin melts in hot weather, blocking the hive entrance and leading to the collapse of nest structures. Others believe that the resin releases fumes that are harmful to bees.

I now doubt that cadaghi is a serious threat to stingless bee hives. Claims about resin melting do not seem to stand up to scrutiny. Giorgio Venturieri placed some hive structures built of cadaghi resin in a container with adults of the stingless bee *T. carbonaria*. He placed it in a controlled temperature cabinet and slowly increased the temperature. At around 44°C, the bees were all dead but the resin was still firm and showing no sign of collapse. At higher temperatures, the resin melted.

Beekeepers who open a hive after it has died may observe dead bees and melted cadaghi resin. They then attribute the death to the resin when heat may be the primary reason the colony died and the resin may have melted later.

Nor does there seem to be any evidence that resin fumes harm colonies of stingless bees. It is possible that *T. carbonaria* can accumulate enough cadaghi seeds around the sticky hive entrance to seriously reduce the size of the entrance and compromise ventilation when it is most needed during the hottest part of the year. But even this does not seem to cause many problems; the bees remove seeds if they do not want them there.

There is no need for you to remove cadaghi resin or seeds from inside the hive or around the entrance, they do no harm.

Cadaghi trees and stingless bees have evolved with each other. The tree benefits from the seed dispersal activities of the bees and has a unique capsule structure to attract bees and allow them to collect resin and disperse seeds. In turn, the bees benefit from the cadaghi resin. It is not in the trees' interest to kill their seed dispersers.

Whether or not this plant poses a threat to hives, it is certainly another fascinating aspect of the biology of these bees.

13.10 The African tulip tree

The nectar and pollen of the introduced African tulip tree (*Spathodea campanulata*) are toxic to insects. The dead bodies of native bees and various other insects inside African tulip flowers that have dropped to the ground bear witness to the poisonous properties of this species (Figure 13-24).

In the flowers of some plant species, alcohol – which is toxic to bees – can form when nectar ferments. But, in the case of the African tulip, it is not alcohol but the toxins in fresh nectar and pollen that kills insects. Flowers of this plant evolved to be pollinated by birds, so the nectar is not toxic to them. The plant may have evolved nectar toxic to insects because the dead insects are an attractive food for birds and so encourage their visits.

African tulip tree is a common ornamental in urban areas of Queensland, the exact zones where bees do extremely well. However, bees do not appear to be strongly attracted to this plant, so it may not pose a great danger. Its overall effect on colonies appears to be minimal. Having said that, we really do not know the effect of African tulip tree on hives, so it would be judicious to minimise contact between bees and this tree. Furthermore, African tulip tree is an environmental weed in Australia, and beekeepers should encourage and participate in control efforts to reduce potential risks to bees and the environment generally.

FIGURE 13-24

Dead stingless bees in the flower of an African tulip tree.

IMAGE
FLAVIA MASSARO

IMAGE **TOBIAS SMITH**

Research on stingless bees has led to improvements in their management.

Here Benjamin Kaluza and the author are investigating the performance of stingless bee colonies in various environments, in this case natural forest.

14 Natural enemies of stingless bees

Many *Bembix* wasps attacking stingless bees at the hive entrance.
IMAGE **GLENBO CRAIG**

Many pests attack stingless bee hives but they can be managed.

The most destructive pests are those that **enter the nest** and feed on the contents (hive syrphid fly, hive phorid fly, small hive beetle).

Hives can be protected from pest entry by good design and construction, and by following good management practices.

Some pests **prey on foraging bees** outside the hive (e.g. *Bembix* wasps, assassin bugs, spiders). These are harder to control but rarely cause major problems.

The stingless bee braconid wasp is a **parasite** that lays its eggs inside bees. This grisly creature is luckily not too common.

A potential **brood pathogen** of hives has been identified, but is currently extremely rare.

Pests of honey bees do not affect stingless bees, with the one exception of small hive beetle. Even the feared varroa mite will not attack stingless bees.

Many different natural enemies attack stingless bee hives, often employing grotesque methods when they do. Parental guidance is recommended for this section. It is the stuff of horror movies … maggots, body snatchers, zombie brain-suckers, and more!

Curiously, stingless bees rarely suffer from diseases caused by microbes, but many insect pests can cause them grief. The most destructive pests are those that enter the nest and feed on the contents (e.g. hive syrphid fly, hive phorid fly, small hive beetle, native hive beetle, pollen mites). Other pests prey on foraging bees outside the hive (e.g. *Bembix* wasps, assassin bugs, spiders). Some are parasites that lay their eggs on the bees (e.g. the stingless bee braconid wasp).

In this section, I describe the weird and grisly habits of these creatures, along with strategies to manage them. Strategies for managing specific pests are discussed under the headings for those pests, while general strategies for enhancing your hive's defences are grouped together in the last section of this chapter, "Protecting your hives from natural enemies".

Note that is it is rarely clear whether hive death was caused by a natural enemy and, if so, which one did it. It is usually quite difficult to diagnose the cause of hive death. By the time a problem is recognised, various natural enemies may be present. Whether one or more of these enemies caused the demise or moved in afterwards is often ambiguous. But a good understanding of these enemies will help you to determine what may be attacking your hive and what you can do about it.

14.1 Hive syrphid fly, *Ceriana ornata*

The hive syrphid fly, a native to Australia, is possibly the most serious pest of stingless bees in our region. It appears to be specialised to use just stingless bee nests as breeding sites. The adult is a fine-looking mimic of a wasp, but the larvae are revolting maggots that lay waste to vulnerable hives (Figure 14-1).

Hive syrphid fly adults are strongly attracted to stingless bee nests and often turn up within minutes of a hive being opened. They are particularly active in hot weather. Adults lay their eggs in crevices around the hive. The larvae hatch, attempt to penetrate the propolis defences that seal gaps and, if successful, start feeding. They begin feeding on honey and/or pollen but end up eating all nutritious organic material in the hive, including the bees, converting everything to a slimy mess.

Adult syrphid flies do not try to enter the hive entrance, so protecting that will not help. Minimising other entry points is the best defence against this pest. This can be done by starting with a well-built hive box and by cleaning the surfaces where the hive sections join before closing the box. (Read more about good building techniques in Chapter 9 "Constructing hive boxes for stingless bees".)

FIGURE 14-1 *Ceriana ornata,* the hive syrphid fly.
A: An adult perched near a stingless bee hive.
B: Eggs laid in the crack between top and bottom sections of a stingless bee hive.
C: Larvae (maggots) destroying a stingless bee hive.
IMAGES **JEFF WILLMER**

14.2 Hive phorid fly, *Dohrniphora trigonae*

Also native to Australia, these small flies can devastate nests of stingless bees. The adult flies can be distinguished by their hunched back (Figure 14-2).

Unlike the hive syrphid fly, the adults of this species invade through the entrance of a hive. They seem to be able to evade the entrance guards. The hives most vulnerable to hive phorid flies are those created by a division that have a new bottom section and an unprotected entrance.

Adult phorid flies probably do not do too much harm, but their larvae feed on nest provisions, particularly pollen. In large numbers, they convert a healthy nest to a disgusting mess in a short time (Figure 14-2). Luckily, Australian phorids are not nearly as serious as the nasty little phorid fly, *Pseudohypocera kerteszi,* which is the biggest killer of stingless bee hives in tropical America.

Traps have been developed to attract and kill phorid flies using a mix of vinegar and sugar water. Dean Haley recommends mixing stingless bee honey, a pinch of pollen and a hint of vinegar until a human can just smell it, then diluting it 50% with water. The traps certainly attract and destroy many adults, but it is not proven whether they reduce the problem.

The best way to control the phorid fly is to reduce points of entry, particularly after a hive division. Help to defend the nest entrance by creating a propolis ring. (Read more about this in "Protecting your hives from natural enemies" below.)

FIGURE 14-2
A: Stingless bee *(LEFT)* compared to adult phorid fly *(RIGHT).*
B: Larvae of the hive phorid fly destroying a nest of *Tetragonula carbonaria.*

14.3 Small hive beetle, *Aethina tumida*

The small hive beetle is native to Africa but was detected in Australia in 2002. It is primarily a pest of honey bees and is, in fact, the only pest of honey bees that also attacks stingless bees. Since its introduction, it has spread widely across the eastern states and destroyed many honey bee hives. (Read about the effect of this pest on honey bees in "Are bees under threat?" in Chapter 15.)

Adult small hive beetles (Figure 14-3) enter the hive and lay eggs. The resulting larvae cause extensive damage by their feeding and by inoculating the hive with a fungus that converts honey comb to a slimy mess (Figure 14-4). Fully grown larvae leave the hive to pupate in the soil. Affected hives are sometimes identified by characteristic stains around the entrance left by departing larvae (Figure 14-5).

The small hive beetle is more attracted to honey bees, and is generally only a minor pest of stingless bee hives. However, in certain areas, at certain times, it can be a real menace to stingless bee hives. Breeding sites include honey bee hives, both wild and managed, so this pest is a greater threat to stingless bee hives in areas close to honey bee hives. Nearby honey bee hives are particularly threatening if they are not managed and are heavily infested with the small hive beetle. The threat is also worse in hot, humid weather.

On the other hand, stingless bees are not totally defenceless against small hive beetles. Mark Greco and colleagues have shown that *T. carbonaria* colonies can defend themselves by applying propolis to the invading adult beetles. Megan Halcroft has shown a similar effective response of hives of *Austroplebeia australis* (Figure 4-24). Megan also showed that worker bees remove the larvae of the small hive beetle. These defence mechanisms are usually effective, but small hive beetles do sometimes enter and destroy hives so, occasionally, further interventions are needed. You will learn more about these in the final section of this chapter.

FIGURE 14-3 A stingless bee *(A. australis)* on the left and three individuals of small hive beetles showing size variation. IMAGE **MEGAN HALCROFT**

FIGURE 14-4
Larvae of small hive beetle in the pollen pots of a *Tetragonula carbonaria* hive *(ABOVE)*.
Larvae of small hive beetle that invaded and destroyed an apparently healthy *T. hockingsi* hive *(ABOVE RIGHT)*.

FIGURE 14-5 The entrance of a hive killed by small hive beetle is often marked by smears left by departing larvae.

Larval pests of hives

Hive syrphid fly	Hive phorid fly	Hive beetle

FIGURE 14-6 Comparison of the larvae of the major pests.

14.4 Which gross grub is that?

The larvae of the three pests that are common in dead or dying hives (hive syrphid fly, hive phorid fly, small hive beetle) are quite different in appearance on close inspection.

Small hive beetle larvae have three pairs of short legs and a brown head capsule, while neither of the fly larvae have legs or a brown head capsule. Syrphid fly larvae have a dark brown disc at the tail end. Phorid fly larvae have protuberances on the side of their body (Figure 14-6).

Fully grown larvae vary in size from small (phorid fly), through medium (small hive beetle), to large (syrphid fly). However, size is not a useful way of distinguishing the species because you won't usually know what stage larvae are at.

Another larva that you may see in a dead hive is that of the American soldier fly, *Hermetia illucens* (Figure 14-7). They are easily distinguished from other larvae in hives because they are flattened, brown and leathery. They commonly breed in decaying organic matter in compost bins. They enter dead hives and do not attack living hives.

FIGURE 14-7
Larvae of the American soldier fly, *Hermetia illucens.*

IMAGE
MELISSA BALLANTYNE

14.5 Native hive beetles, *Brachypeplus* spp.

Species of native hive beetles in the genus *Brachypeplus* also invade stingless bee colonies. I don't consider them an enemy of stingless bees, but I include them here because you may see them in your hive and be concerned about their presence.

Native hive beetles are in the same family (Nitidulidae) as the small hive beetle, but are smaller, more slender, and have parallel sides and short wing cases (Figure 14-8, Figure 14-9). These native beetles rarely seem to be a pest. Occasionally they are seen in great numbers in a dead hive but probably did not kill it. They seem to generally only eat detritus in the hive and cause no problems at all.

FIGURE 14-8 Native hive beetles. From the left, two larvae, two adults from above, adult from below, and a stingless bee. IMAGE **MEGAN HALCROFT**

FIGURE 14-9
Adult native hive beetle *Brachypeplus*.
IMAGE **KEN WALKER**

14.6 Pollen mites, *Cerophagopsis trigona* and other mites

Mites only seem to appear in numbers when a hive has died, and probably do not trigger the death of the hive. The presence of mites is marked by fine grains at the base of the hive or in pollen pots (Figure 14-10), which appear to be the mites' accumulated excrement. As in the case of the native hive beetle, I do not consider them a significant enemy.

The feared varroa mite pest of honey bees will not affect stingless bee colonies. We are confident of this because this mite has co-existed with many species of native stingless bees for thousands of years in Southern Asia, with no reported cases.

FIGURE 14-10
Mites have converted the pollen in this pot to dust *(ABOVE)*.

Close-up of mites *(RIGHT)*.

FIGURE 14-11
Drawing of the underside of female and male pollen mites, *Cerophagopsis trigona (BELOW)*.

ARTWORK **A FAIN**

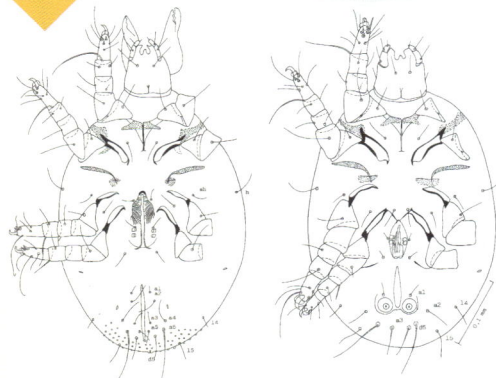

FIGURE 14-12 A female *Bembix* wasp digging a hole for its nest; observe how she uses her front legs to fling the sand grains backwards. IMAGE **BERNHARD JACOBI**

14.7 *Bembix* wasps

Bembix are one of the group known as "sand wasps" because of their habit of digging their nests in sand ((Figure 14-12). These wasps excavate a tunnel and build a series of cells. They stock each cell with prey and then lay an egg in the cell so that the hatching larva can feed on the prey.

The prey of *Bembix* are well known, thanks to the thorough studies of Howard Evans. Most *Bembix* species catch flies for their young. However, one group of Australian native *Bembix* specialises in preying on bees. The three small, closely related species in this group are widespread and common in Australia.

Bembix tuberculiventris preys on solitary bees in southern Australia but, in the north, preys on stingless bees, both *Tetragonula* and *Austroplebeia*. This wasp species feeds mainly on males but will also take workers. It probably hunts at flowers.

Bembix flavipes, in three widely spaced localities in northern Australia, feeds only on *Austroplebeia* males. It hunts at stingless bee nest entrances.

Bembix musca occurs all the way down the east cost of Queensland and New South Wales. It is an extreme specialist, as its prey consists entirely of *Tetragonula* bees, nearly all males. It hunts at stingless bee nest entrances (Figure 14-13).

Another species, not so closely related, is *Bembix moma*. This is another widely distributed species. *B. moma* is not as strict in its prey type, but mainly takes bees. In northern Australia, prey includes stingless bees, all males.

It is common to see these wasp predators flying around a hive entrance attempting to intercept bees (Page 196). You may also see adult female wasps digging nests in sandy soil, particularly in areas that are partly shaded (Figure 14-12). The cell at the end of their burrow typically contains about 25 adult bees and a wasp egg or larva consuming them.

As the wasps prey mainly on male bees, they do not seriously threaten colonies. It is intriguing to contemplate why they do only take males and thus limit themselves to a small proportion of a nest's bees. Perhaps they are avoiding the strong mandibles of the female workers?

14.8 Stingless bee braconid wasp
Syntretus trigonaphagus

While some wasps are predators (e.g. the *Bembix* species discussed in the previous section), many others are parasites.

The *Syntretus trigonaphagus* wasp is a **parasitoid**, a particular kind of parasite that lays its young in a host. The young eventually develop to a size similar to the host and then emerge, killing the host in the process (Figure 14-14, Figure 14-15).

This stingless bee parasitoid belongs to a very large group of wasp parasitoids called braconid wasps. Generally, braconids attack the larval stages of their host, but this one attacks the adult. *Syntretus trigonaphagus* is native to Australia, but a close relative, *Syntretus splendidus*, attacks adult bumble bees in Europe.

The adult wasp stalks a foraging bee on a flower or at the nest entrance. When it makes contact, it lays an egg inside the adult bee. The larva hatches, feeds and grows inside the bee's abdomen. When it reaches its full size, the larva emerges from the bee's rear end. Cunningly, it waits until the bee has left the nest so that when the larva emerges, it falls to the ground where it can safely pupate. Finally, the adult wasp emerges from the soil and attempts to repeat the life cycle.

The host range and geographic range of this wasp are not known. It seems to attack only species of *Tetragonula* and not *Austroplebeia*. It geographic range probably mirrors that of its *Tetragonula* hosts.

FIGURE 14-15 The stingless bee braconid wasp parasitoid *Syntretus trigonaphagus*. **A:** An adult laying an egg in its bee host, which is foraging on a flower. IMAGE **JOHN KLUMPP**
B: Side view of female. IMAGE **MALCOLM RICKETTS**

14.9 Minor predators

14.9.1 ANTS

Ants are not generally a problem for stingless bees in Australia if hives are solid and without cracks or entry points. This is different from the Americas, where ants regularly destroy strong and established stingless bee hives. Australian stingless bees defend the hive entrance and ventilation hole very effectively and will stop ants entering by biting them or daubing them with propolis.

Some beekeepers apply a ring of grease, bowl of water or other ant deterrent around the hive mount. Indeed, I even suggested this myself when I wrote my first paper on propagating hives in 1988. But I now believe that it is an unnecessary precaution, because ants quickly learn to keep away from hives or risk a deadly defensive reaction. The exception to this may be after a hive transfer or split, when large gaps prevail between hive sections.

Ants will also destroy natural nests where a tree falls down and splits open. If ants gain access to the colony through these gaps, they can quickly do real damage.

FIGURE 14-14 The stingless bee braconid wasp parasitoid *Syntretus trigonaphagus*. **A:** Mature parasitoid larva inside the abdomen of its *T. carbonaria* host. **B:** Larva partially emerging from the abdomen. **C:** Larva fully emerged from the abdomen of its host. IMAGES **MALCOLM RICKETTS**

14.9.2 ASSASSIN BUGS

Assassin bugs are common around hives, particularly on farms. They ambush bees at the hive entrance, pierce them with their stout mouthparts, and inject saliva, which liquefies the bees' body contents. They then suck the resulting juice out of the bees (Figure 14-16). Assassin bugs will also defend themselves by biting humans. Their saliva causes a painful bite so they should not be handled. Use a flyswat to eliminate assassin bugs.

FIGURE 14-16 An assassin bug feeding on a bee near the hive entrance.

14.9.3 ASIAN HOUSE GECKOS AND CANE TOADS

I am occasionally asked whether Asian house geckos and cane toads, both introduced species that are now common in many of Australia's warmer parts, pose any threat to stingless bee hives.

Generally, Asian house geckos and cane toads should not be a major problem for your hives. Both species are nocturnal, so tend not to meet stingless bees, which stay entirely inside their nest at night. In contrast, toads can attack honey bee hives because honey bees leave their nests on hot nights, becoming susceptible to these tough predators.

Occasionally, toads may be active at the entrance of a stingless bee nest during the day. If you think this is possible, elevate the hive by at least 30 cm. (Elevation also helps the bees to take flight, especially in winter.)

Geckos may shelter in niches around your hive.

If the hive is in a dark position, geckos may be drawn to the nest entrance during the day. I have seen them feeding on bees at a hive that was on the rafters of an outdoor garage. Ensure your hive is not in a position where you think geckos could be active during the day.

14.9.4 SPIDERS

Spiders may build webs around a hive or ambush bees at flowers. In particular, crab spiders are specialists at ambushing flower-visiting insects (Figure 14-17). Interestingly, studies have shown that stingless bees can detect and avoid flowers with crab spiders.

You can do little about these ambush predators on flowers, but you can keep the area around your hive free of spiders and webs, especially the area in front of the hive where most bee traffic concentrates.

Before opening a hive, use a brush to remove any surrounding spider webs. When bees fly out of the nest to defend themselves they disperse more widely than usual and can become entangled in webs if any are present.

FIGURE 14-17
This crab spider has successfully ambushed a stingless bee on a flower. IMAGE **DANIEL KLAER**

14.10 Pathogens

Globally and in Australia, confirmed cases of brood pathogen attack on stingless bee hives are rare. Perhaps pathogens are held at bay by the antimicrobial properties of the propolis. (Read more about this in "Building materials: wax, resin and propolis" in Chapter 4.) Infections may also be kept in check by stingless bees' single use of brood cells. (Read more about this in "Brood rearing in social bees" in Chapter 4.) Pathogens of honey bees benefit from the reuse of cells, which allows pathogen propagules to infect the next larva reared in the cell.

However, evidence of a possible brood infection is emerging. Jenny Shanks found and documented the first known brood disease of stingless bees in Australia, in New South Wales in 2014. I call this "Shanks' brood disease" (with her approval!).

Jenny isolated and identified a strain of *Lysinibacillus sphaericus* bacterium as the causative agent. Initial signs of infection consist of brown discoloured larvae either in brood cells (Figure 14-18) or deposited on surrounding hive structures (Figure 14-19); healthy larvae are white. As the infection develops, the larvae lose their characteristic features and become unrecognisable, fluid-filled sacs. Eventually, the number of cells in the brood chamber decreases (Figure 14-20), affecting the adult population. All known infected colonies have ultimately died. So far, one strain of bacterium has been confirmed in *T. carbonaria*, while a different strain has been verified in *A. australis* colonies. The disease does not appear to be causing many problems and has been only rarely observed by Queensland beekeepers, but it certainly needs to be monitored.

I suggest that you watch for discoloured brood, especially in hives that are not gaining weight and are inactive. If you suspect you have a brood disease, take photos and store a sample of brood in the freezer.

If this disease becomes prevalent, we will need to change some of our beekeeping practices to reduce the risk of spread. It may be a good idea to adjust procedures now in anticipation of future problems. For example, you could consider wearing disposable gloves, sterilise tools, and improve general hygiene when working on more

FIGURE 14-18 Brood cells infected with brood bacteria show discoloured larvae dispersed among normal white larvae. IMAGE **JENNY SHANKS**

FIGURE 14-20 Healthy brood (*LEFT*) and unhealthy brood of *Tetragonula carbonaria* infected with the brood bacteria (*RIGHT*). IMAGE **JENNY SHANKS**

FIGURE 14-19 Larvae infected with the brood bacteria lose their form and may be removed from their cells by workers. IMAGE **JENNY SHANKS**

than one hive to reduce the possible spread of any pathogens between colonies. You could also cease practices that involve transferring brood and food between hives, and destroy diseased hives by burning. However, I must point out that I have not yet adopted any of these practices because I have never seen any of the disease symptoms in hives that I manage, and the cost and inconvenience would be considerable.

As yet, there are no documented fungal or viral brood infections in stingless bee colonies. Denis Anderson looked for viruses in *T. carbonaria* and found that only one colony in 24 contained virion-like particles, which proved not to be any of the common honey bee viruses. He was surprised at how rare viruses were in *T. carbonaria* compared to honey bees.

14.11 Protecting your hives from natural enemies

14.11.1 ENHANCING DEFENCES AGAINST PESTS THAT ENTER HIVES

The worst pests are those that enter the hive. Your colony is most susceptible after a hive manipulation (e.g. division, honey extraction). The process of colony transfer creates a particularly great danger.

For pests that enter hives, the best defence is to prevent entry. Points of entry include gaps between hive sections, and the hive's structural holes (entrance hole etc). To reduce gaps, ensure you have a well-built box that has flat joining surfaces between the hive sections. (Read more about this in Chapter 9, "Constructing hives boxes for stingless bees".)

When opening boxes for division, honey extraction, etc., clean any spilled honey from the surfaces and scrape the surfaces to remove any propolis or other material that may prevent the sections coming together tightly. Bees seal gaps as quickly as they can, so help them by ensuring these gaps are narrow. If you see wide gaps after closing the hive, then seal them with tape.

Recently transferred or divided hives are most susceptible to invasion by natural enemies. Of the two hives produced by splitting, the one with the original top half is particularly susceptible. This is because this half has a new bottom with

an entrance that is unprotected. Eventually, the bees will build a tube internal to the entrance. This tube, approximately 150 mm long, is a key feature of nest defence. Guards line this gallery and fiercely repel any enemies attempting to enter. Without the tube, it's easier for pests to sneak in. It takes the colony days or even weeks to build this tube. This may be too late for a hive if the intensity of pest attack is high.

One control option is to split in cooler seasons when pests are less active. You can also seal entrance holes if you expect high pest pressure. You can totally seal the holes after a transfer or division with a breathable closure for a few days (Figure 9-10). In this time, the bees can concentrate on rebuilding their defences in the absence of pests. Open the hole after a few days to allow the bees to remove wastes and begin foraging. If the hive needs to re-queen, the newly selected virgin will need to leave the hive for her mating flight. By then, the colony will have repaired damaged food pots, cleaned up spilled food, sealed gaps, rebuilt entrance tubes and be able to protect themselves.

Alternatively, cover the entrances with a mesh screen to allow the foraging bees to continue working. Some keepers of stingless bees who keep hives in areas that suffer from small hive beetle use this technique. Most hive beetles are larger than stingless bees but their size can vary a lot (Figure 14-3). If the apertures are the correct size (approx. 2.5mm), bees will be able to move through the screen and all but the smallest beetles will be kept out (Figure 14-20). The mesh is a temporary measure for a few weeks only, until the bees have built up their own defences. The mesh screen technique will only work for the small hive beetle, as it is the only pest that enters through the entrance and is larger than the bees.

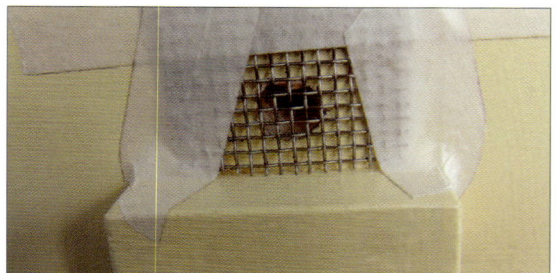

FIGURE 14-20 A hive entrance covered with mesh to deter small hive beetles.

If you are a busy beekeeper, and are unable to return to your hive to unseal it, you can apply a partial seal of propolis. I recommend this as the most practical and effective solution to preventing pest entry after a colony manipulation. Make a propolis ring, press it firmly around the entrance, and squeeze it to reduce the entrance size to about 5 mm diameter (Figure 14-21). Do the same on the inside of the entrance hole. The propolis rings increase the length of the entry and reduce the diameter at two points, helping the guard bees to repel natural enemies. Propolis can be obtained at any opportunity when a hive is opened. It should be processed before use. (Read more about this in the feature "Propolis from stingless bee hives".)

FIGURE 14-21
A propolis ring used to reduce the size of the entrance hole in preparation for a colony transfer or division.

14.11.2 RESCUING AN INFESTED COLONY

In the extreme case of the external defences failing and pests entering the hive, then you may still be able to save the colony by moving the brood into another hive box. The internal nest pests (hive syrphid fly, hive phorid fly, small hive beetle and native hive beetle) all start by consuming stored food. As they grow, their ability to move and consume the brood increases. If you can catch them before they enter the brood, you may be able to save it.

Have a new, clean hive box ready. Place a propolis ring around the entrance. Open the infested hive and identify the brood chamber. Using a knife and your hands, separate the brood from the rest of the nest. Place only the brood, without any food pots, into the new box. The loss of its food stores will be a major setback for the colony but it may permit its survival. It is crucial to try to get as many workers as possible into the new box. Some will have come with the brood. These are important house bees that will continue to care for the brood. But the colony is without stores and so also needs foragers to bring back fresh

food. Use an aspirator ("pooter") to collect bees and release them into the new hive.

Leave the infested hive open to make it less acceptable to the bees inside, which may then leave and look for the new nest. Move the infested hive to one side and put the new hive in the original position so that foragers return to the new hive.

Eventually, you should destroy the infested nest contents so that the pests do not complete their life cycle and threaten other hives. Scrape the contents out onto scrap paper, wrap them up, and double wrap the resulting bundle in plastic bags. This can go directly into the bin. Alternatively, if you want to be really sure that the pests will not escape the bin, freeze the bag for a few days or, if the weather is hot, leave it in the sun.

14.11.3 PROTECTION FROM PESTS THAT OPERATE OUTSIDE THE HIVE

Pests that operate outside the nest are unlikely to seriously threaten a colony, but you can take action to reduce the number of bees lost to these pests. Use a flyswat to kill adult hive syrphid flies, assassin bugs, spiders etc. Wipe away any spider webs, particularly those in the flight path of the foraging bees. It may also help to eliminate any spaces around the hive where pests can find refuge. For example, geckos, spiders, etc. may shelter in spaces under roofs and in polystyrene insulating covers.

14.11.4 RE-USING A DEAD HIVE

If your colony of bees dies, you can clean and re-use the hive box.

Once you have disposed of the infested nest contents, scrape out any remaining contents with a hive tool or paint scraper. Do not worry about removing the last traces of propolis from inside the box. Then scrub the sections with a stiff brush in plenty of water in a tub. Sit them with their open sides down to drain. Complete the drying and disinfect the sections by placing them open side up in the sun for a few days. Then store them in a well-ventilated but dry place, ready to use.

It is a good idea to repaint the outside of the hive at this time. Fill any holes or gaps and apply a coat of acrylic exterior paint following the directions on the can. Wait at least a week for the paint to off-gas before re-using your hive.

This final part tackles the critical issue of pollination, and what role bees play in pollinating both wild and cultivated plants.

You'll find out how pollination works, whether there really is a global decline in bee populations, and how **you** can support the insects that support our food production.

We'll also talk about whether stingless bees could play a greater role in food pollination in Australia, and which crops are likely to benefit most from stingless bees as pollinators.

Finally, you'll find a wealth of handy tips on using stingless bees for pollination on farms.

Part THREE

Bees for pollination

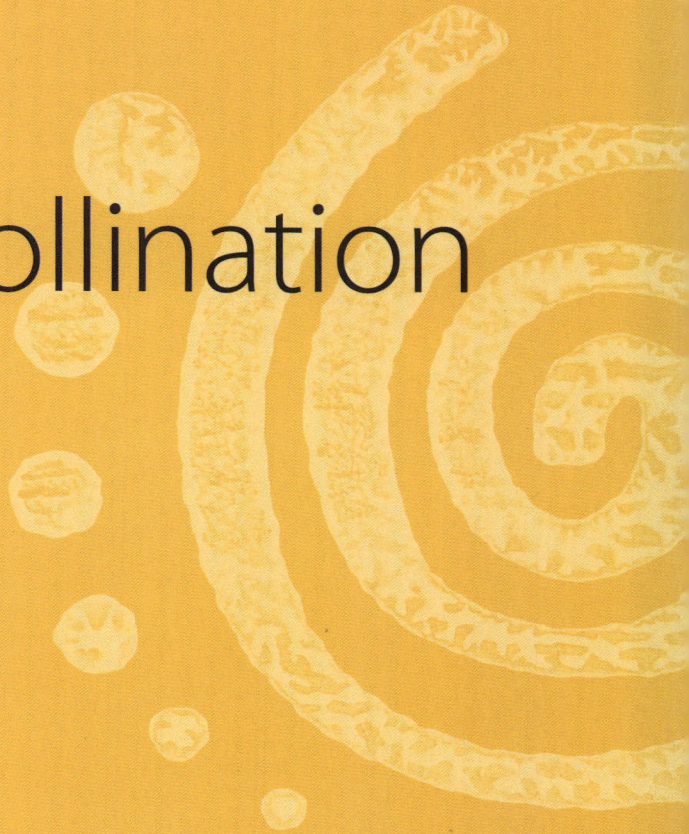

Tetragonula carbonaria visiting a strawberry flower. IMAGE **TOBIAS SMITH**

An Amazonian stingless bee (*Melipona fasciculata*) visiting a capsicum flower. IMAGE **GIORGIO VENTURIERI**

15 Bees and Pollination

Pollination is crucial for the health and survival of natural plant communities, and is mega-important for agricultural production.

Pollination **moves pollen** from the male to the female part of a flower, where it fertilises the plant's ovules, allowing seed production.

The European honey bee is the most important managed pollinator, but **wild bees and other insects** are often more valuable for transferring pollen. A diversity of bees is optimal for maximum crop yield.

Honey bees and wild bees face the **threats of habitat loss,** climate change, pesticide use and the spread of their natural enemies, so we need to manage them prudently.

We can **act now to help pollinators** by managing our landscapes, providing nest sites, and taking care when using pesticides.

We can also **propagate** certain pollinator species to boost their populations.

FAST FACTS

The health of bees in our natural ecosystems and on farms is currently the subject of much concern. Furthermore, there are increasing calls to diversify our crop pollinators and to manage native bees to promote food production.

What is the evidence for these concerns and suggestions? Are our bees really in crisis? Are our crops at risk of failing for want of pollination? If so, why bother with native bees when it's easier to continue to put our faith in honey bees? If it is justified to use native bees, then what is the best way of using them? The answers to these questions and more lie in this chapter.

15.1 The importance of insect pollination

15.1.1 WHAT IS POLLINATION?

Pollination is a fundamental process in natural plant communities, and in agriculture.

Pollination is the process by which male sex cells (pollen) move from a flower's **anthers** to its **stigma** to enable fertilisation of the ovule, which lies beneath the stigma within the ovary (Figure 15-1). The fertilised ovule then produces seeds (this is called **seed set**). The pollen cannot move on its own, and requires an external agent of some kind to achieve the transfer.

The transfer of pollen can be within or between flowers on the same plant (**self-pollination**), or from one plant to another of the same species (**cross-pollination**). Cross-pollination is especially important. In wild flora, crossing prevents inbreeding and creates a next generation of plants that are genetically diverse, vigorous and able to cope with a changing climate. In agriculture, cross-pollination boosts the yield and quality of many crops. It is not about maintaining the genetic diversity of plants; it's about producing more of the produce that humans need.

Crops are a simplified version of nature. Individual plants of a crop variety (e.g. Hass avocado) are genetically identical; they are clones of each other. Movement of pollen between these plants is equivalent to self-pollination. Varieties are often in rows. If cross-pollination is required, then pollen needs to travel from a row of one variety to a row of another (Figure 15-2). It is crucial then that bees move across rows, and not just along them.

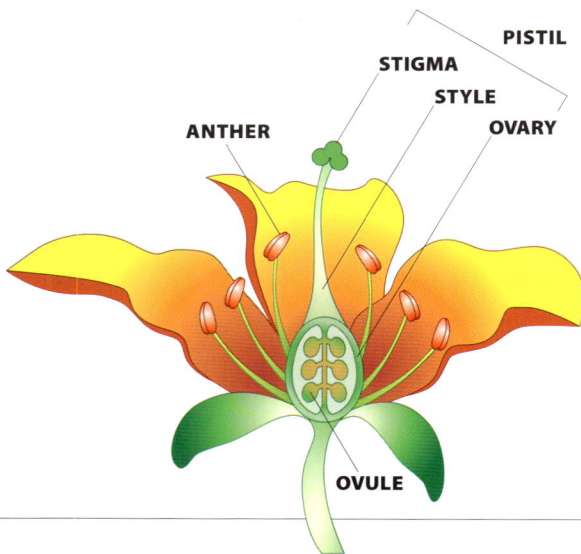

FIGURE 15-1 A typical flower. The pollen is produced in the anthers but must be moved to the stigma by an external agent. Once in place, the pollen grain germinates and grows down the style into the ovaries where it fertilises the ovules. These will grow into seeds.

FIGURE 15-2 Cross-pollination and different types of self-pollination in a tree crop. Trees of the same variety are genetically identical, so pollination between trees of the same variety is also said to be self-pollination. ARTWORK **GINA CRANSON**

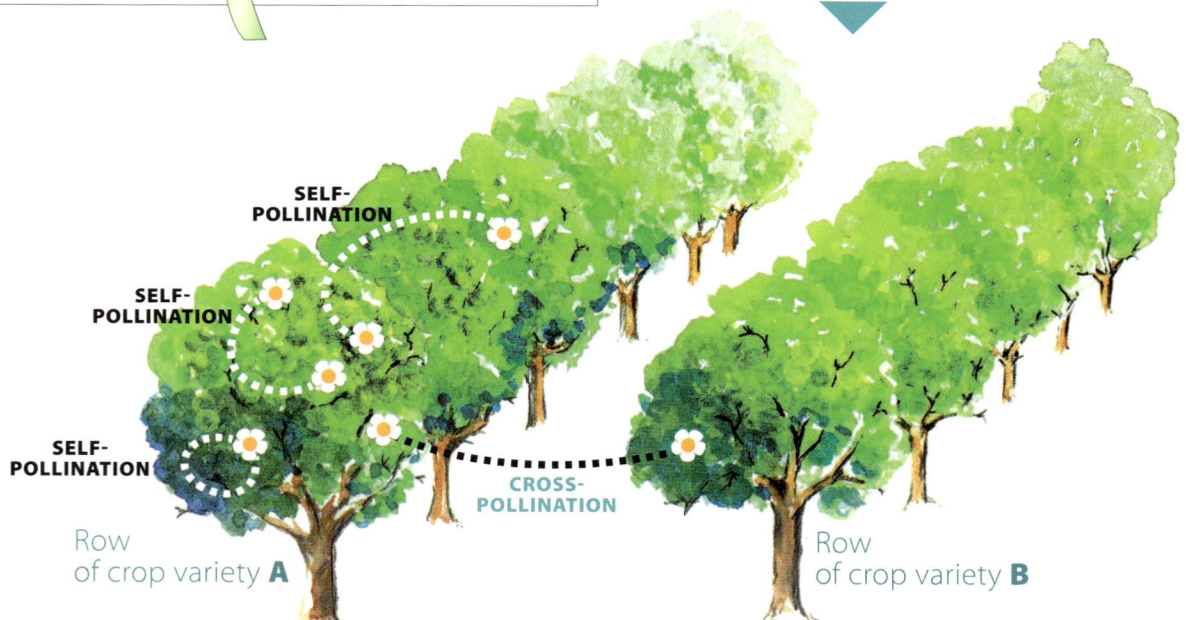

Pollen may be moved by wind or water (abiotic pollination) or by animals (biotic pollination). Important biotic pollinators are winged animals such as birds, bats and insects. Many animals play worthy roles, but bees are paramount in terms of abundance and effectiveness. Bee pollinators have evolved in parallel with flowering plants since flowering plants appeared in the early Cretaceous period about 120 million years ago.

The coevolution of plants and pollinators over millions of years has resulted in the most fantastic diversity of flower shapes, structures, sizes and colours — and an equally intriguing range of pollinator behaviours. Most plants lure pollinators with nectar, and then trick them into transporting pollen as the price for a free drink. Some flowers form a long tube that is only accessible with a long tongue. And some flowers need to be vibrated before they release their pollen. (Read more about this in the section on buzz pollination in "Some dazzling examples of native solitary bees" in Chapter 2.)

Why did some flowers evolve biological chastity belts that require a particular "key" to achieve fertilisation? Why not do as many flowers do, and make the pollen freely available to all flower visitors? The answer is simple, really. Restricting access to pollen to certain, specialised pollinators gives the flower an advantage. It encourages bee visitors to be "faithful" to the flowers of that species, meaning more cross-pollination and better seed set.

The evolution of the flower and biotic pollination changed the face of our planet, as flowering plants diversified and dispersed to dominate most plant communities. Today, pollination remains a crucial process in the health of natural ecosystems, and the health of bees continues to be central to effective pollination.

15.1.2 HOW MUCH DO HUMANS DEPEND ON POLLINATION?

Biotic pollination is of immense importance to the agricultural systems that produce much of our food. The annual global value of animal pollination has been estimated at $100–200 billion dollars. Yields of about 75% of our crop species are improved by animal pollination.

Dependence on pollinators varies. Some crops are 100% dependent, meaning they will not set any seed unless insects visit their flowers. Others will yield but at a lower level without pollinators to transfer their pollen. In Australia alone, we grow 1,000,000 ha of crops that yield best only in the presence of pollinators, and the value of the increased yields is around $5 billion. Economically, pollination is a mega-important process.

On the other hand, some assertions about the importance of bee pollination are exaggerated. The most objective studies show that only about 5% of total annual world food production would be lost in the absence of insect pollinators. This is because the staple crops that provide the bulk of the human diet are grains and roots, which do *not* need animal pollination. The grains come from grasses such as wheat, rice and corn, which are wind-pollinated. Grasses are also the primary food stock of the animals that we eat. Root crops, such as potatoes and cassava, are produced vegetatively and do not need flowers, insects or their interactions to produce food.

Nevertheless, many valuable crops do benefit from insects visiting their flowers. Examples of crops that depend on bee pollination for more than 90% of production are sunflower, almond, apple, blueberry, cherry, lucerne, white clover, pumpkin, zucchini, rockmelon and watermelon. Even coffee yields are increased when bees visit coffee flowers, so don't forget to thank the bees for your next cup of coffee!

Here are some crop categories that yield best when pollinated by bees.

- **Oil seeds** such as sunflower and canola
- **Fruit crops** such as blueberry and melons
- **Vegetables** such as pumpkins
- **Nut crops** such as almonds
- **Protein crops** such as beans
- **Stimulant crops**, in particular coffee
- **Animal food** such as lucerne and clover
- **Biofuel crops** such as Jatropha

Pollination is important for directly producing food, and also key to producing seeds to grow the next generation of crops. Even vegetables such as carrots, which develop from a root, need pollination to set the seed needed to sow next year's plants. Ditto for forage crops such as clover, where the product is leafy material but the seed is still vital to producing more plants.

15.1.3 WHICH ARE THE MOST IMPORTANT POLLINATORS?

The European honey bee is the most important *managed* pollinator — that is, the most important pollinator introduced by humans. European honey bees are proven pollinators of canola, sunflowers, almonds, apples, stone fruit, kiwifruit and cucurbits.

But other *wild* insects are probably of even greater consequence in increasing crop yields through pollination (Figure 15-3). Wild bees have often proven to be more valuable than managed honey bees. Insects other than bees can also be central to pollination. Papaya (or "pawpaw", as we know it in Australia) is pollinated by hawkmoths; annona fruits (such as the custard apple of Australia) are pollinated by pollen beetles; and cacao (for chocolate production) is pollinated by tiny, non-biting midges. All these insects are wild species that breed naturally within agricultural ecosystems. Their contribution is a valuable ecosystem service that humans receive from nature. In addition, as we shall learn later in this chapter, their populations can be augmented by sound environmental management.

FIGURE 15-3 Wild bees foraging on crops, such as this one visiting a lychee flower for pollen, may increase farm yields. IMAGE **GIORGIO VENTURIERI**

A recent international survey of pollinators in 41 crop systems worldwide, by Lucas Garibaldi and his colleagues, showed that wild insects pollinated crops more effectively than honey bees. An increase in wild insect visitation enhanced fruit set by *twice as much* as an equivalent increase in honey bee visitation. The survey also showed that visitation by wild insects and honey bees promoted fruit set synergistically (Figure 15-4). That is, honey bees alone increase yields, wild pollinators alone increase yields, and both types of bees in combination provoke even higher yields. So, to enhance global crop yields, we need to manage both honey bees and diverse communities of wild insects.

How do these combinations of honey bees and wild insects boost yields? Evidence from a broad range of crops in many habitats indicates that the diversity of visitors is just as important as the overall abundance. So, for example, 10 visits to a flower by 10 different insects will result in better fruit set than the same number of visits by just one insect species.

Synergism among the insect visitors ensures better crop yields. This is because different pollinators may pollinate flowers on different parts of the plant, or work on days of varying weather, or perhaps work in different parts of the field. Pollinators of different species may also interact with each other in a manner that is positive for pollination. For example, when stingless bees encounter honey bees, both individuals are likely to flee the scene; that movement sends them to another plant, which receives the pollen recently collected. The end result of a diverse pollinator assemblage is better coverage and better flow of pollen to where it does most to increase yield.

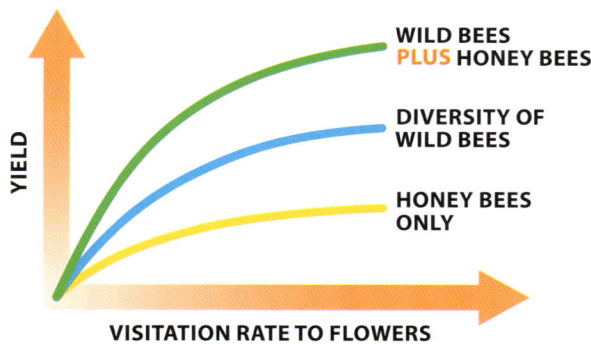

FIGURE 15-4 Crop yields are increased by visits from honey bees, but increase more following visits by a diversity of wild bees. Honey bees and wild bees in combination increase yields the most.

Our natural ecosystems that provide the wild insect pollinators are not separate but deeply interwoven with our agricultural landscapes. It is crucial that we conserve these areas. The biodiversity of our native ecosystems provides a kind of biological insurance, a guarantee of crop pollination even if managed bees are removed.

15.2 Are bees under threat?

We have heard much in recent times about the decline in bee populations worldwide. Here I consider whether managed honey bees and wild bees are really in crisis, as we hear regularly in the media.

15.2.1 HONEY BEES

Honey bees *are* in decline in some parts of the world, but not everywhere. For example, there has been a 60% decline in honey bee hives in the USA since the 1940s. A serious decline occurred with the introduction of the varroa mite in the 1980s. Since 2006, the decline in the USA and Europe has been accelerated by Colony Collapse Disorder, a syndrome in which many worker bees disappear from a hive.

But strong growth in honey bee numbers in Eastern Europe, Asia, Latin America and Africa means that the number of managed honey bee hives worldwide has actually *increased* in recent years. Where declines have occurred, the most likely causes are pesticide use, climate change, habitat loss, and new diseases and parasites.

Honey bees are still in reasonable shape in Australia, possibly because less intensive land use means they enjoy better habitat and less exposure to chemicals. Typical landscapes in Australia include areas of eucalypt forest, which provides abundant resources for bees. But the introduction of the small hive beetle has wiped out many honey bee hives. Furthermore, if the varroa mite gains a foothold in mainland Australia, as is feared, it is likely to decimate feral and managed colonies.

15.2.2 WILD BEES

Wild bees also appear to be in decline in the parts of the world where their numbers have been monitored (e.g. western Europe). In Australia, so few studies have been conducted that we have little idea how our wild bees are faring.

Globally, a revealing study by Rachael Winfree and colleagues has shown how human disturbance of ecosystems has affected wild bees. This meta-analysis (a statistical analysis of multiple previous studies) shows that bee abundance and species richness are both reduced by disturbance, but not by a huge amount; wild bees are still in pretty good shape. But further declines are likely, given the forecasts of continuing conversion of natural areas to farms (Figure 15-5). The disturbance types that cause the greatest decline in bee populations are 1) habitat loss and fragmentation, 2) loss of forage, 3) loss of breeding sites, and 4) population subdivision (where bee populations are separated from each other, and suffer from inbreeding).

I explore the major threats to honey bees and wild insects in more detail below.

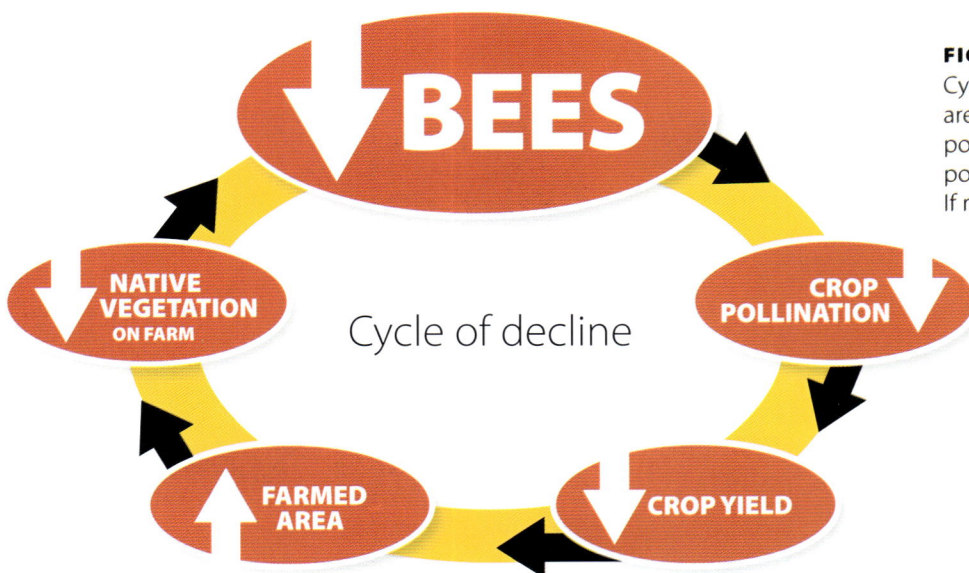

FIGURE 15-5
Cycle of decline of natural areas of native vegetation, pollinator abundance, crop pollination and crop yield. If reduction in yield causes more land clearing, the situation spirals downwards.

ARTWORK
GLENBO CRAIG

15.2.3 HABITAT LOSS

Many agricultural landscapes around the world continue to be cleared of remaining natural vegetation. This intensification of land use is taking place in both developed and less developed countries. The loss of remaining forest, vacant land, hedgerows, etc. has a negative impact on wild insects and honey bees alike. Bees are affected by a loss of alternative forage, loss of nest sites, loss of landscape continuity and loss of genetic diversity. The loss of alternative forage may be a particularly serious risk. Not only does this forage provide continuity of food for bees, but a diversity of pollen in the diet of many bees provides superior nutrition and maintains good health.

15.2.4 PESTICIDE USE

Worldwide, agricultural industries are trying to reduce pesticide use, but control of pest insects, fungal pathogens, etc. still relies on these chemicals. Pesticides are a double-edged sword — on the one hand, they control pests but, on the other, they reduce the abundance of beneficial insects.

One group of chemicals called the **neonicotinoids** is currently under close scrutiny. Originally considered to be relatively benign to bees, they are now thought to cause serious sub-lethal effects. Bees are not killed by these chemicals, but their nervous system is affected so that they may, for example, become disoriented and unable to find their way back to the hive. Pressure is increasing to restrict the use of these chemicals, but the evidence is still not strong enough to galvanise policymakers into banning them completely. Some European nations have placed a moratorium on their use.

This is a complex situation, because banning this class of chemicals may result in alternatives being used for pest control that are actually worse for the environment.

15.2.5 DISEASES AND PARASITES OF BEES

Honey bees suffer from numerous diseases and parasites. Some have spread from other bee species, while others have exploded from an originally isolated distribution to occupy vast areas of the globe.

Below I discuss two examples of new pests: the varroa mite and the small hive beetle. We know a lot about the pests and diseases of honey bees, but all organisms, including wild bees, have their natural enemies. (Read more about this in Chapter 14 "Natural enemies of stingless bees".)

The varroa mite (*Varroa destructor*)

The varroa mite is an external parasite that sucks the haemolymph (blood) of adult and immature bees in the hive (Figure 15-6). The varroa mite is a threat only to *Apis* honey bees; it breeds only in their hives and will not attack other bee species. In fact, in mainland Asia, *Varroa destructor* is a natural harmless parasite of the Asian honey bee (*Apis cerana*).

Some races of the mite transferred to the European honey bee when humans moved the more utilitarian European species to Asia, into the native range of the Asian honey bee. The mite soon developed an ability to infest and produce offspring on the European honey bee, which enabled it to spread throughout the world, except to Australia, which is now the only continent free of varroa mites.

European honey bees have little natural resistance to these pests and their colonies are devastated by mite invasions. In addition to feasting on bee blood, the varroa mite transmits viruses. If varroa does establish and spread in Australia, the effect on agricultural pollination could be even greater than in Europe or North America. Much crop pollination in Australia is provided by feral honey bees. Feral honey bees would be most affected by the varroa mite because managed hives can be treated for the pest.

FIGURE 15-6 Varroa mite on a honey bee pupa. The pupa has been pulled from its cell to show the mite. IMAGE **DENIS ANDERSON**

The small hive beetle (Aethina tumida)

Although Australia is currently free of the varroa mite, apiarists are struggling with another pest and this one affects both honey bees and stingless bees: the small hive beetle.

This insect was first found in Australia in 2002. It is called the small hive beetle because, in its native range of southern Africa, larger hive beetles also exist. It causes loss of hives and increases the cost of managing hives. Adult beetles (Figure 15-7) enter bee hives. If the hive is not strong and the beetle numbers are high, the bees are unable to repel the beetles, which remain and lay eggs. The beetle larvae then attack the honey comb by eating through it and adding microbes that infest the honey, rendering it slimy and unacceptable to bees and humans. Fully grown larvae (Figure 15-7) leave the hive to pupate in the soil.

Small hive beetles are worse in some areas than others and also worse in hot, humid weather. The small hive beetle has killed many managed honey bee hives on the Australian east coast, and there is anecdotal evidence that feral honey bee colonies have also been decimated. Professional beekeepers manage the pest largely by using pesticides, chemicals that can also kill bees and, therefore, need careful application and management.

FIGURE 15-7
Small hive beetle in a honey bee hive, adults *(ABOVE)* and larvae *(LEFT)*.
IMAGES **DENIS ANDERSON**

15.3 Protecting and enhancing the pollinators

It is clear that bee pollinators are in decline in key areas and face further reductions unless we take action. Bee losses are being caused by habitat loss and fragmentation, pesticide use, and epidemics of pests and diseases. These threats degrade the free pollination service that our natural and agricultural systems receive. So, what can we do about it?

Pollinator declines can be countered by landscape management practices and by the more direct means of developing managed pollinators.

15.3.1 MANAGING THE LANDSCAPE TO ENHANCE POPULATIONS OF WILD BEES

We can manage our landscapes to protect and enhance the pollinators. Many actions can improve pollinator abundance and diversity in our natural and production systems. These actions need to address each of the major threats (Table 15-1). Some examples of actions that successfully address the threats follow.

TABLE 15-1 Threats to pollinator abundance, and diversity and actions to address the threats.

Threat	Action needed to remedy threat
Habitat loss — general effects	Increase available bee forage and nest sites by retaining pollinator habitat
Habitat loss — loss of bee forage	Plant for pollen and nectar supply
Habitat loss — loss of breeding sites	Provide artificial nesting sites
Habitat loss — fragmentation impedes pollinator movement	Retain and re-establish corridors to connect habitats and increase colonisation
Habitat loss — fragmentation causes pollinator population subdivision	Retain and re-establish corridors to allow pollinators to migrate and mix genes
Insecticide poisoning	Use insecticides judiciously
Pests and diseases	Maintain border security Breed bees for resistance Minimise spread of disease

FIGURE 15-9 The owner of this macadamia farm has planted native vegetation in low-lying areas to provide forage for pollinators.

FIGURE 15-8 Natural vegetation *(BACKGROUND)* next to this blueberry farm *(FOREGROUND)* in northern New South Wales supports populations of stingless bees that visit crop flowers in prodigious numbers.

15.3.2 INCREASE AVAILABLE BEE FORAGE AND NEST SITES BY RETAINING POLLINATOR HABITAT

The benefits of retaining remnant vegetation on farms were quantified by Taylor Ricketts and colleagues on a coffee farm in Costa Rica. They found that wild bee pollinators from remnant rainforest fragments on the farm increased coffee yields by 20%. This increase translated into extra revenue of US$60,000 per year. This value exceeds revenue from any other use of the land. So protecting the forest remnants yielded double benefits: for biodiversity and for agriculture. Australian farms often include or border on areas of natural vegetation that provide valuable pollinators (Figure 15-8).

15.3.3 PLANT FOR POLLEN AND NECTAR SUPPLY

Some astute farmers are planting vegetation on their farms to provide food sources for bees (Figure 15-9). Species are selected that meet a number of criteria, the foremost being that they flower over a long period and are favoured sources of food for bees. Native perennial species provide many environmental benefits, but annual exotics may also help to provide a quick feed for local bees, other pollinators and beneficial organisms.

15.3.4 PROVIDE ARTIFICIAL NESTING SITES

Providing artificial nesting sites for bees can deliver great benefits for farmers. For example, farmers in some tropical countries provide and protect nesting timber for carpenter bees for pollination of passionfruit. Bee hotels, in which a variety of nest substrates are provided, are also becoming common in urban areas, conservation areas and on farms (Figure 15-10).

15.3.5 RETAIN AND RE-ESTABLISH CORRIDORS

Corridors of natural vegetation that connect larger wild areas are crucial elements in healthy human landscapes. They aid dispersal of plants and migration of animals, including pollinators. Pollinators need to move to find food and nest resources, and corridors facilitate this movement. This boosts the abundance of pollinators and encourages their movement to the landscapes where they are desired for pollination. Corridors also prevent the isolation of insect populations and

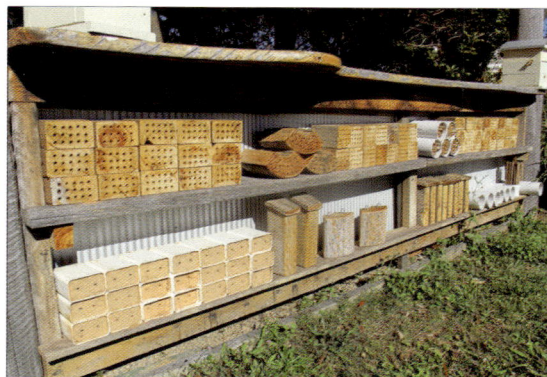

FIGURE 15-10 Bee hotel to provide nest sites for wild bees, this one by Chris Fuller in Kin Kin, Queensland. IMAGE **GLENBO CRAIG**

subsequent inbreeding. Inbred populations can decline and eventually become locally extinct. For flying insects, corridors do not necessarily have to be continuous strips; even closely spaced patches will allow migration.

15.3.6 USE INSECTICIDES JUDICIOUSLY

The negative impacts of insecticides are lessened by careful use. Before using them, first consider more environmentally friendly options such as biological and cultural control. It is also advisable to monitor pest species, and spray insecticides only when absolutely necessary.

If you do have to spray, select the insecticides that are least toxic to bees, and avoid spraying when the crop is flowering. Spray at night when pollinators are absent, and spray when the air is still and drift is reduced. (Read more about this in "How to minimise the effects of pesticides" in Chapter 16.)

15.4 Direct management of pollinating insects

In the previous section, I discussed how we can manage our landscapes to help the wild bees that breed naturally and visit our crops to forage on flowers and pollinate them. Here, I consider bee species that can be directly propagated by humans.

In addition to honey bees, about 10 other bee species are being managed this way around the globe. Notable examples include alfalfa leaf-cutter bees for lucerne in North America, orchard bees for apples in Japan, and bumble bees for pollination of glasshouse crops in Europe. The European example is pertinent to Australia and worth expanding upon.

Bumble bees are used to pollinate glasshouse-grown tomatoes, and other crops, in many parts of the world. When these crops are grown in fields, wind will usually provide enough vibration to allow tomato flowers to self-pollinate but, in the still conditions of a glasshouse, another agent is required. It is a tribute to human ingenuity that colonies of the bumble bee *Bombus terrestris* are now propagated in enormous numbers and supplied to the greenhouse tomato industry.

Bumble bees are not native to Australia and are only present as an introduced species in Tasmania. Despite efforts from the greenhouse industry to gain permission to introduce

them, this has not been allowed by Australian authorities, with good reason. In parts of the world where the bumble bee has been introduced, it has a record of escaping from greenhouses and invading the natural environment, with negative consequences for other bee species.

The Australian tomato industry currently depends on mechanical vibrators for pollination, but this is very costly (Figure 15-11). A better solution is needed.

FIGURE 15-11 Manually pollinating tomato flowers using a rod vibrator. IMAGE WITH PERMISSION OF **SOILLESS AUSTRALIA**

Katja Hogendoorn and colleagues have shown that Australia's native solitary blue-banded bees (*Amegilla* spp.) and carpenter bees (*Xylocopa* spp.) can pollinate tomatoes in greenhouses very well, and that it is possible to rear blue-banded bees continuously in these conditions. But impediments prevent their adoption by industry just yet.

The major hitches are:

- plastics used in greenhouses, which filter out UV light used by the bees
- the difficulty of preventing bees from escaping through greenhouse vents
- the need to use insecticides, which can kill the bees
- a disease that attacks the brood of blue-banded bees
- maintaining a good-quality food supply
- a tendency of these bees to enter diapause (suspended development) and remain inactive in cooler seasons.

Further work is needed before blue-banded bees work as well as bumble bees for greenhouse pollination, but investment in this research could yield long-term and substantial returns. In the next chapter, we will discuss how we can manage our native stingless bees for crop pollination.

A stingless bee hive well positioned on a strawberry farm.

16 Stingless bees for crop pollination

FAST FACTS

Stingless bees boast a number of traits that make them valuable crop pollinators.

Stingless bees possess some advantages over honey bees, but also a number of shortcomings for farm use.

Six crops are promising targets for pollination by stingless bees in Australia: macadamia, mango, avocado, lychee, blueberry and strawberry.

Using stingless bees for crop pollination is new, but useful guidelines are provided here.

Hives can be moved onto farms just for the flowering season or can remain there for the whole year.

Hives can be strategically placed to maximise their effectiveness, take advantage of the best microclimate, minimise fighting, and minimise exposure to insecticides.

Solid progress is being made on using stingless bees for crop pollination. The possibility of employing these bees in farm management on a large scale now seems real.

In this section, I explore the role of stingless bees as crop pollinators, with an emphasis on Australia. I discuss stingless bees' strengths and weaknesses for this mission, and nominate a number of crops that I think would be good targets. I also identify the species of stingless bees that I believe are our best options. I finish with a section on how to manage these bees on farms.

If you have not read Chapter 5 "Foraging behaviour of stingless bees", now is a good time to do so. It will build your knowledge base and aid your understanding of this chapter.

16.1 What makes stingless bees useful pollinators?

Stingless bees in general possess a number of traits that make them valuable as pollinators.

These traits are listed below. Note that not all species of stingless bees demonstrate the same biology, so not all of the following attributes apply to all species. Also, most of these qualities are held in common with honey bees, but distinct from other bees, such as primitively social bumble bees and solitary bees. (The special advantages that stingless bees hold over honey bees appear in the next list.)

- Most species of stingless bees are **generalists** and opportunists. Colonies are able to exploit the pollen and nectar of many plant species. They readily adapt to introduced exotic plant species, including crop species. The ability of a bee species to use the pollen of many plant species is called **polylecty**.

- The generalisation at the colony level described in the previous point is tempered by **floral constancy** at the level of the individual bee. **Floral constancy** means that a forager on a particular trip usually only visits one plant species.

- Colonies can be kept in **hives**. Concentrating bees in a hive makes it easier to manage a large number of individuals simultaneously. Management may include inspection, propagation, feeding, re-queening, managing pests, transporting, shielding from insecticides and protecting from extreme weather.

- Colonies of stingless bees are **perennial**. There is no need to breed new colonies each year (as is the case for bumble bees) and workers forage continuously, across seasons, within constraints imposed by temperature and light.

- Large **reserves of food** (pollen and honey) are stored in nests. This has the obvious benefit of allowing colonies to survive long periods of low food availability. Additionally, it means that workers will collect floral resources beyond their immediate needs, resulting in focused visits to preferred flowers.

- Workers **recruit** nestmates to rewarding floral resources and provide information on the position of those floral resources, which allows the rapid deployment of many foragers. (Read more about this in "Recruitment communication of foragers to resources" in Chapter 5.)

- Stingless bees are able to forage effectively in **greenhouses**. For example, Megan Halcroft found that *Austroplebeia australis* adapted readily to the confines of the greenhouse chamber, visited the crop flowers and significantly increased the seed set in celery and carrot crops.

16.2 Unique advantages of stingless bees over European honey bees

Stingless bees also possess some advantages over honey bees.

Stingless bees are generally harmless to humans and domesticated animals. The sting of honey bees can be painful, or even dangerous to allergic individuals.

Stingless bees support conservation efforts because their propagation helps to conserve our natural biodiversity.

Stingless bees are naturally present in many warmer areas of Australia and in the tropics broadly.

Stingless bees thrive in hot and wet climates where European honey bees may not do well. (The flip side is that they are not suited to cooler climates where honey bees thrive.) An exception is African races of honey bees, which are well adapted to tropical conditions and thrive in their native Africa and as introduced species in the American tropics.

Stingless bee colonies are unable to abscond (abandon their hive). Some races and species of honey bees regularly leave their nest and move to another location. The queen of a stingless bee colony is unable to fly again after her mating flight so her colony is fixed in its original position.

Stingless bees are resistant to the diseases and parasites of honey bees so they provide a much-needed "insurance policy" for pollination. The many viral, bacterial, fungal and microsporidian diseases of honey bees do not affect stingless bees. Even the much feared varroa mite is specific to honey bees. While stingless bees have their own suite of natural enemies, some of which can be serious, they are distinct from those of honey bees. Honey bees and stingless bees are thus complementary.

✓ **Stingless bees do not need access to water**, unlike honey bees. Honey bees need to be provided with a water supply in hot weather to enable them to regulate temperatures in their hive.

✓ **Stingless bees have short flight ranges** relative to honey bees. A range of approximately 500 m is typical for small Australian species, compared to honey bees' 5 km range. This shorter flight range means stingless bees are more likely to confine themselves to the target crop. The vast area and potentially greater number of plant species within the flight range of honey bees may see them attracted to non-target species.

✓ **Stingless bees can help to control crop pests** by bolstering the local ecosystem. Frank Adcock, a macadamia grower and keeper of stingless bees, has noticed that foraging bees can provide a constant supply of prey to creatures such as birds, frogs, lizards and beneficial insects. These predators may then help to control crop pests. This strategy may allow reduced insecticide applications, with further benefits for pollinators and natural enemies of insect pests.

✓ **Certain species of stingless bees can buzz pollinate**, which is vital for some plant species. Honey bees and most stingless bees cannot buzz pollinate. But one group of tropical American stingless bees called *Melipona* can (Figure 16-1).

16.3 Weaknesses of stingless bees as crop pollinators

While stingless bees have many strengths as pollinators, they also have a number of shortcomings that relate to their ecoclimatic tolerances, to their innate behaviours, or to our lack of experience and knowledge in managing their hives.

● **Climatic limitations.** Generally, stingless bees are limited to warmer parts of the globe, so they do *not* offer options for crop pollination in temperate zones. Other climatic limitations may also apply; for example, the Australian species *Austroplebeia australis* is easily kept in drier areas of inland Australia, but bees of this species lose weight and may even die when they are moved outside of their preferred habitat.

● **Fighting and usurping.** Colonies of *Tetragonula* species may attempt to usurp other colonies, particularly when they are placed in the artificially close proximity required for pollination.

● **Low availability of hives**. The keeping of stingless bees has boomed in the last two decades, but we are still years away from being able to provide the large numbers of hives needed for commercial-scale crop pollination. One threat to availability is the competing demand for the hives, driven by interest from various sectors including city-dwellers who enjoy stingless bees as pets, garden pollinators, and for production of coveted "sugarbag" honey. This high demand has kept hive prices relatively high, which may inhibit adoption by farmers. In the future, hive prices are expected to drop as their supply increases exponentially.

● **Slow growth rates.** Stingless bees may show slow colony growth rates when compared to honey bees. Advances in hive management are overcoming this limitation.

● **Small size.** Stingless bees are generally smaller than honey bees and thus may transfer less pollen. They also visit fewer flowers per unit of time and so pollinate at a lower rate than honey bees and other large bees.

FIGURE 16-1 An Amazonian stingless bee *(Melipona fasciculata)* buzz pollinating a tomato flower.
IMAGE **GIORGIO VENTURIERI**

- **Damaging plants.** In Central and South America, some stingless bee species damage plants to produce resin flows for them to collect. Fortunately, none of our Australian species have been observed to do this.

- **Robbing flowers.** Some stingless bees will extract pollen or nectar resources and may even damage flowers without effectively pollinating them. This depends on the bee–flower interaction. For example, stingless bees in Brazil have been shown to rob passionfruit flowers (Figure 16-2). However, honey bees and many other types of bees are equally guilty of this offence.

FIGURE 16-2 A stingless bee robbing a passionfruit flower by drilling a hole and taking nectar without pollinating. IMAGE **GIORGIO VENTURIERI**

16.4 Crops likely to benefit from stingless bees in Australia

Worldwide, 18 crop species are reported to be pollinated by stingless bees. Six crops are the most promising targets in Australia: macadamia, mango, avocado, lychee, blueberry and strawberry.

Other possibilities exist, but they are either minor crops in Australia (e.g. coffee, rambutan, longan, coconut and guava), their pollinator requirements are uncertain (e.g. citrus), they are not visited or effectively pollinated by stingless bees (e.g. cucurbits, such as melons and pumpkins), or they are planted across broad acreage and require large numbers of hives in areas where colony populations are low (e.g. sunflower and canola).

The six nominated crops are prime targets because they meet the following criteria:

- crop yield and/or quality benefit from pollination
- crop flowers are effectively pollinated by stingless bees
- stingless bees are naturally attracted to the crop
- flowering occurs at a time of year when the bees are active.

In the main production areas of subtropical east coast of Australia, these crops mostly flower synchronously in winter to spring. This is not ideal. Sequential flowering of one crop after the other would allow movement of hives from one farm to the other, but the opportunities to do this are limited.

16.4.1 CROP FIDELITY

"Crop fidelity" refers to the proportion of bees from a hive that forages only on the target crop in which the hive is placed.

Giorgio Venturieri, Chris Fuller and I recently determined the crop fidelity of colonies of two stingless bee species on five crops in subtropical Australia: macadamia, avocado, lychee, blueberry and strawberry.

On average, 52% of pollen foragers from introduced hives of these two species of stingless bees visited only the crops in which they were placed (Figure 16-3). This varied between plant species, being highest on macadamia (85%) and lowest on blueberry (12%).

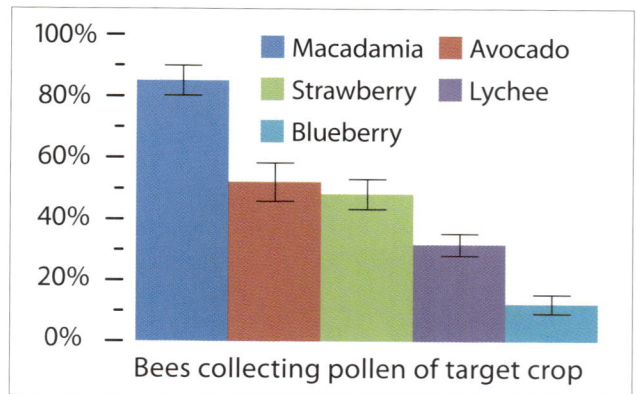

FIGURE 16-3 The proportion of bees collecting pollen from the target crop as opposed to other plants, when their hive was placed in that crop.

MACADAMIA

FIGURE 16-4 *Tetragonula* stingless bee visiting a macadamia flower.
IMAGE LEFT **JEFF WILLMER**

16.4.2 MACADAMIA

Both the quantity and quality of the nuts produced in a macadamia crop increase when bees visit the flowers. The inter-planting of different varieties of macadamia trees that are compatible with each other, so that cross-pollination can occur, also benefits yields. Insects aid cross-pollination by moving pollen from trees of one variety to another.

Many insects visit the flowers of macadamia trees. The most common visitors in the main production areas in Australia are honey bees and stingless bees (Figure 16-4). Both types of bee are usually present for the entire macadamia flowering season from about August to September.

Native stingless bees and commercial honey bees generally behave differently on macadamia flowers. Honey bees collect mainly nectar from the base of the flowers, while stingless bees collect mainly pollen. This brings stingless bees into closer contact with the stigma, the small surface on top of the flower on which the deposited pollen grains grow and eventually fertilise the flower.

Stingless bees are proven efficient pollinators of macadamia trees. When macadamia flowers are enclosed in bags so that they cannot be touched by insects, they set no nuts. When flowers are enclosed in cages that exclude honey bees but allow visits by the smaller stingless bees, they yield as many nuts as flowers that are not enclosed at all, showing that stingless bees alone are efficient pollinators of this crop.

When honey bee hives are introduced onto farms, only 24% of pollen-collecting bees in the colonies forage on macadamia flowers, while the remainder prefer the flowers of other plants. The equivalent figure for stingless bees is 85% (Figure 16-3). Stingless bees respond well to heavy flowering of macadamia, becoming much busier and taking advantage of the temporary abundance of pollen. Stingless bees also forage in the darker inner canopy of macadamias to pollinate flowers where honey bees don't go. However, honey bees can fly at lower temperatures than stingless bees, so they forage for an average of 10 hours per day compared with only 7 hours per day for stingless bees.

Stingless bees are common on farms that have surrounding natural vegetation, mainly eucalyptus forest. The natural vegetation provides nest sites and other food for bees. The larger the area of natural vegetation around a farm, the greater the bee population. However, even in some areas with only 15% remnant vegetation cover, adequate bee populations are present. Stingless bees *are* absent from some farms where most natural surrounding vegetation has been cleared. In those areas, it could be particularly beneficial for a beekeeper to introduce boxed stingless bee hives.

FIGURE 16-5 *Tetragonula* stingless bee visiting a mango flower. IMAGE **GIORGIO VENTURIERI**

FIGURE 16-6 *Tetragonula* stingless bee visiting an avocado flower. IMAGE **GIORGIO VENTURIERI**

16.4.3 MANGO

Visits by insects to mango flowers increase yields. The simple flowers of this plant allow pollination by most visiting insects.

Studies in Brazil, India, and Australia have revealed that stingless bees are the most common insect visitors to mango trees (Figure 16-5). But this is only part of the equation. To be an effective pollinator of a particular plant, an insect needs to do more than just visit. Denis Anderson showed that stingless bees visiting mango flowers in Australia carried the largest amount of pollen on their bodies, made close contact with the stigma, and left the most pollen grains on the female part of the flower. They were, therefore, the most efficient pollinators of mangoes.

Furthermore, stingless bees moved more frequently from tree to tree and thus were probably the most effective cross-pollinators. Honey bees are not strongly attracted to mango flowers and are only occasionally observed around them. Flies are common visitors to mango flowers in many parts of the tropics, and are also efficient pollinators.

16.4.4 AVOCADO

The rich buttery flavour of avocado is brought to you courtesy of a small tree and its pollinating insects. The pollination of this crop is often carried out by honey bees. In lands as far-flung as South Africa, Israel, Jamaica and Australia, honey bees are used to ensure that avocado flowers are fertilised.

But this marriage between avocado trees and honey bees is a new arrangement. The avocado tree comes from the jungles of Central America. Honey bees do not naturally occur in the New World. Today, honey bees are used because they are the only species available in sufficient numbers.

How well does this marriage of convenience work? Not particularly well. Honey bees do not really like visiting avocado flowers, only visiting when there is nothing better flowering nearby. How effective are they at pollinating avocado? Here they do score a little better. Individual bees collect pollen quite effectively, but those collecting nectar don't do a very good job.

So, do other insects exist that are likely to be better pollinators of avocadoes? In short, yes. Recent studies in Mexico have shown that many local species of stingless bees are frequent visitors and efficient pollinators of avocado flowers. These bees, so common and abundant in tropical America, were probably the original pollinators of this plant before it was taken out of the forests and cultivated. In Australia, local stingless bees are moderately common visitors to flowering avocado trees (Figure 16-6). It is

LYCHEE

FIGURE 16-7 *Tetragonula* stingless bee visiting a lychee flower. IMAGE **GIORGIO VENTURIERI**

BLUEBERRY

FIGURE 16-8 *Tetragonula* stingless bee visiting a blueberry flower.

not proven, but I believe that they are probably just as effective avocado pollinators as their Central American relatives.

Certain aspects of avocado production in Australia make it a particularly good target for stingless bees. First, trees are often well spaced and so allow good light penetration. Second, production is concentrated in areas of Queensland where stingless bees are active. Third, flowering occurs in late spring, a time of year when the bees are active. Fourth, insecticides are rarely used during flowering.

On the other hand, these advantages need to be tempered by the fact that Australian stingless bees do not show a strong attraction to avocado flowers. Research is needed to confirm the importance and management of Australian stingless bees in avocado pollination.

16.4.5 LYCHEE

Stingless bees and honey bees are the most common visitors to lychee flowers in India. In Australia, stingless bees (*Tetragonula* spp.) are also common visitors (Figure 16-7). One study showed that stingless bees collected pollen and nectar from lychee flowers but did not often touch the stigma. In contrast, the larger honey bees usually touched the stigmas while visiting the flowers, suggesting that they are more effective pollinators. More observations are needed to determine whether this happens consistently.

The related crops longan and rambutan are also candidates for stingless bee pollination in Australia. Rambutans are effectively pollinated by stingless bees in Brazil.

16.4.6 BLUEBERRY

Blueberry farming in Australia occurs in temperate areas using northern highbush varieties, and in warmer areas using southern highbush and rabbit-eye varieties. In the warmer areas of northern New South Wales, stingless bees are common visitors to blueberry flowers (Figure 16-8).

Stingless bees are particularly common where natural vegetation abuts a blueberry farm (Figure 15-8). Blueberry fruits are larger if flowers are insect-pollinated. The inter-planting of different varieties of blueberry also increases fruit set and size, because it promotes cross-pollination between the varieties.

Pollen from highbush blueberry is sticky and heavy. Many solitary bee species vibrate the blueberry's anthers, causing the pollen to be released. Honey bees and stingless bees are not able to perform buzz pollination, so their effectiveness in pollinating this variety of blueberry is uncertain. There is a pressing need to determine the effectiveness of both bee types and their management in this rapidly expanding industry.

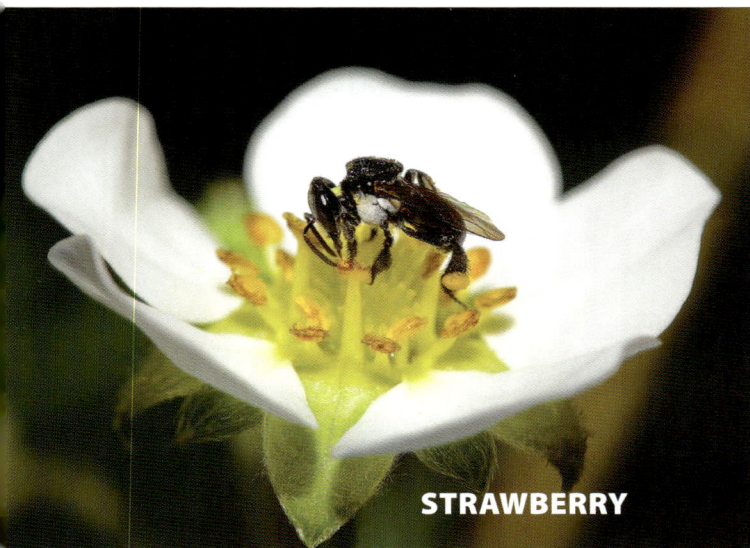

FIGURE 16-9 *Tetragonula carbonaria* visiting a strawberry flower. IMAGE **TOBIAS SMITH**

16.4.7 STRAWBERRY

Strawberry cultivars vary in their need for pollination. Some bear well without insect visitors, while other varieties produce much more fruit of higher quality when flowers are visited.

Imported stingless bees have been evaluated in Japan for pollination of strawberries in glasshouses, where they performed better than honey bees. In Brazil, a stingless bee similar in size to Australian stingless bees was an excellent pollinator of two varieties of strawberry in greenhouse conditions.

Other insects are also able to pollinate strawberry flowers. The centre of strawberry production in Australia is the Sunshine Coast, where stingless bees are naturally abundant (Figure 16-9). In this area, strawberries flower over the cooler part of the year. Local *Tetragonula carbonaria* bees are active in winter when conditions are sunny, but not during periods of cloudy or exceptionally cool weather. A combination of both stingless bees and honey bees is recommended in this situation.

Strawberry plants produce flowers over a long period, which overlaps with fruit production. Insecticides are often used to protect fruits from pests. This can affect pollinators, which are exposed to the insecticides while foraging on flowers.

16.5 Which species of stingless bees are most suitable for pollination?

The best species for pollination are those that meet or exceed the following criteria.

● They adapt to artificial hives. Most Australian species have been moved into hives at some time, and those that do best in hives have become the most common. In particular, *Tetragonula carbonaria*, *T. hockingsi* and *Austroplebeia australis* have adapted well to hives and become the most commonly kept. These three species are possibly also the most common in nature. Their colony mortality rates are low and they are easy to propagate. All species replace the queen naturally and independently following colony division.

● They adapt to environments disturbed by humans. Again, *T. carbonaria*, *T. hockingsi* and *A. australis* perform very well in urban areas and farms.

● They are well adapted to the climate. *T. carbonaria* and *T. hockingsi* naturally occur on the tropical and subtropical east coast of Australia and coexist with major areas of horticultural production. *T. carbonaria* has a more subtropical distribution and *T. hockingsi* is more tropical, so the two species combined can cover much of northern Australia's horticulture. *Austroplebeia australis* occurs naturally in the less humid inland parts of Australia. The bee could be developed and used for broadacre crops such as sunflower and canola, which are cultivated in that area.

● They are active throughout the year in their natural geographic range, although at the southern end of the range, their activity in cooler weather is restricted. Hives of *T. carbonaria* begin foraging activity above 18°C. Hives of *T. hockingsi* begin foraging activity above 20°C. Hives of *A. australis* begin foraging activity above 22°C.

● They form large colonies with many foragers. Hives of *T. carbonaria* and *T. hockingsi* house populations of approximately 10,000 adult bees, of which at least 1,000 are foragers that leave the hive to collect resources on flowers. *A. australis* usually has less than half this population size and is more variable in its foraging activity, but can mobilise lots of foragers to good resources.

- They are not defensive toward humans, although they bite moderately when hives are opened. They are, however, able to defend themselves effectively against natural enemies.

- They have a broad diet range. All stingless bees are more or less generalists, but *T. carbonaria* and *T. hockingsi* are particularly broad in their diet range. *A. australis* has a narrower range.

- Colonies recruit nestmates to rich floral resources. Experimental evidence proves that *T. carbonaria* does this. Probably *T. hockingsi* and *A. australis* can too.

- They can be held at high densities. A weakness of *T. carbonaria* and *T. hockingsi* is that they can expend much energy in fighting activity associated with attempts to usurp the nests of other colonies. *Austroplebeia australis* do not possess this negative attribute.

The two species, *T. carbonaria* and *T. hockingsi*, are outstanding candidates as crop pollinators in Australia. *Austroplebeia australis* has strengths as a pollinator, but is limited by its smaller colony size, slower colony reproduction rates, higher temperature threshold for foraging, unsuitability for humid areas where higher-value crops are grown, narrower diet range, weaker foraging activity, and limited forager recruitment strategies. Despite these limitations, *A. australis* may prove to be effective in drier inland areas.

16.6 Managing stingless bee hives on farms

Honey bees have long been integrated into agricultural systems, but using native stingless bees on farms is new. In this section, I summarise the current strategies used for this novel endeavour and provide a guide to their practical management.

16.6.1 STRATEGIES FOR KEEPING STINGLESS BEES ON A FARM

Stingless bee hives may be owned and managed by the farmer or by a beekeeper. Some farmers find this a personally rewarding side activity, but others prefer to subcontract to an expert beekeeper. If the farm is a poor environment and the bees lose weight and health while positioned there, the beekeeper may require payment from the farmer. If the farmer perceives a benefit in the form of high crop yields, then they may be pleased to pay. In the case where both parties benefit from the arrangement, then both parties may agree that no money needs to change hands.

If the farmer pays the beekeeper, then they may require proof that the hives are strong pollination units. Honey bees can be checked by opening hives and scrutinising the number of frames containing brood and the number of frames covered in bees. The presence of brood stimulates foraging, resulting in pollination of the visited flowers. It is not as simple to check stingless bees. But one can determine the strength of a stingless bee hive by three main factors: weight, foraging activity and brood volume. For pollination, foraging activity is the most important factor. (For how to evaluate each of these factors, see "How to determine the strength of a stingless bee hive" in Chapter 13.)

Hives can be placed permanently or temporarily on a farm. **Permanent placement** of stingless bee hives near target crops on farms is gaining popularity in Australia. In the case of permanent placement, the farmer usually buys the hives. The farmer may care for them or may make an agreement with a beekeeper to manage them. Permanent hives work particularly well in more heterogeneous environments that provide year-round resources. An agricultural ecosystem that consists of a mix of multiple crop species, natural bushland and ornamental garden plant species is particularly favourable. (Read more about this in "Optimal environments for keeping hives" in Chapter 8.) In areas that are less than ideal, it will help to feed the hives. (Read more about this in "Feeding bees" in Chapter 13.) Permanent hives are well suited to organic farms, where there is no threat of pesticide poisoning.

In the case of **temporary placement** of hives on farms, the beekeeper agrees with the farmer to move the hives to the farm for a specified period during the flowering season. The advantage of this strategy is that the hives can be moved off-farm to more favourable locations when the crop is not in flower. The strategy is especially effective where the flowering period is short.

FIGURE 16-10 Stingless bee hives on a macadamia farm. Note hives are located at the ends of rows for solar warming (macadamias flower in subtropical Australia during a cool time of year) and widely separated to avoid drift fighting. IMAGE **GIORGIO VENTURIERI**

Farms with little plant diversity and heavy insecticide use constitute hostile environments for bees. Homogeneous monocultures provide few alternative food sources. Beekeepers may be able to increase their opportunities by moving the hives into alternative crops in succession.

A disadvantage of temporary placement is the high cost of moving hives. This is especially the case with stingless bees, which need to be spaced more widely than honey bees and so demand more time and labour (Figure 16-10). (Read more about strategies for safely transporting hives in "Moving your hive" in Chapter 8.)

Most hives temporarily relocated into farms will normally be from local sources, but there may be opportunities to move colonies long distances. The opportunities for profitable migratory pollination services with stingless bees still need to be identified.

The movement of hives from subtropical Australia to temperate regions may be justified when a temperate crop flowers at a time of year when the bees will be active. But let's consider almonds as one discouraging example. The almond industry is concerned about a shortage of bees to pollinate the increasing area of crop in temperate Australia. But this deficit clearly cannot be met by stingless bees, as almond flowering occurs when the temperature is too low for active foraging by these tropical insects.

Regardless of the strategy for keeping stingless bees on the farm, **regular inspection and maintenance** are crucial. Dead or dying hives need to be removed and decontaminated to prevent spread of pests. Hives that reach full weight need to be divided to build hive numbers. Hives may need to be moved or closed when pesticides are applied. Boxes may need maintenance such as re-painting or even replacing if serious decay renders them unsuitable homes for the bee colony. Hives also need to be securely mounted and protected. (Read more about this in "Hive mounts and stands" in Chapter 9.)

16.6.2 HOW TO DISTRIBUTE STINGLESS BEE HIVES ON A FARM

Stingless bee hives may be distributed evenly across a farm or concentrated at particular sites. Honey bees are typically concentrated at a site called an apiary or bee yard (Figure 16-11). A site of multiple stingless bee hives is called a meliponary, stingless bee apiary, or stingless bee yard (Figure 16-12).

The keeping of stingless bee hives and honey bees in the same yard does not present any problems. They are compatible, even complementary. The only common pest problem is the small hive beetle, so any outbreak needs to be managed to halt spread. The two bee species may compete for the same resources, which could diminish colony performance if they are held at high densities, especially during poor seasons. But their joint presence may help pollination, because diversity of pollinators can be more important than their abundance for increasing yield.

The logistical advantage of concentrating hives spatially is compelling: all hives can be placed

FIGURE 16-12 Stands for mounting many stingless bee hives in Brazil. IMAGE **GIORGIO VENTURIERI**

in one operation at one or a few easily accessible points. But this may not be optimal. For an example of a superior spacing for stingless bees, see Figure 16-13.

FIGURE 16-11 An apiary of closely placed honey bee hives on a macadamia farm. IMAGE **TOBIAS SMITH**

Native vegetation bordering creek – source of wild bees and alternative bee forage

Remnant forest – source of wild stingless bees, conserve it!

Creek

House block

Dam

Apiary of honey bees

Creek

Rows of trees with varieties interspersed

Row of windbreak trees

■ ■ ■ = Native stingless bee hives

Keep stingless bees in the house block (but not honey bees)

Honey bee hives can be kept together but stingless bee hives should be spread out

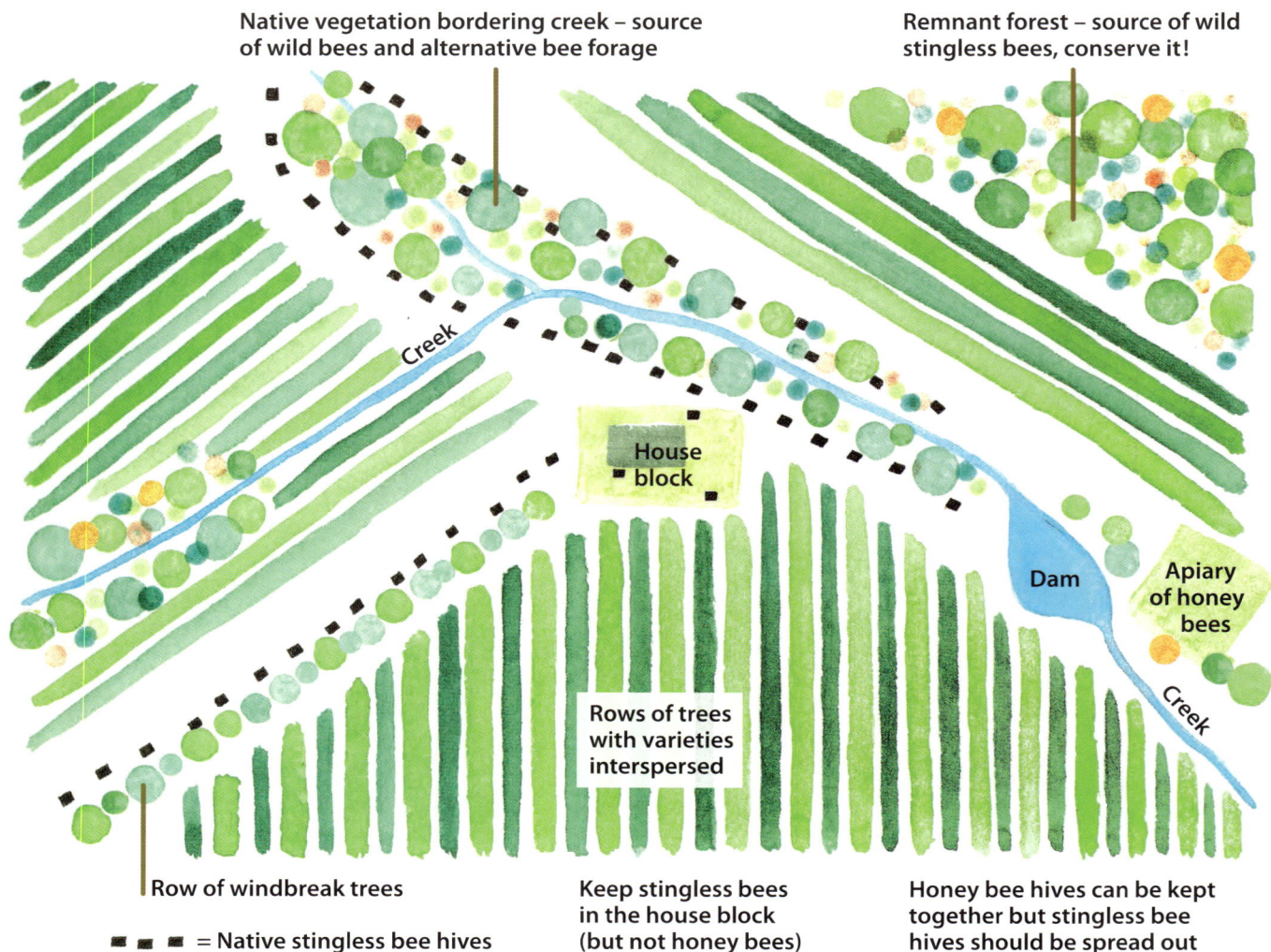

FIGURE 16-13 Stingless bee hives (shown on this map by the square symbol) distributed on a farm. ARTWORK **GINA CRANSON**

Factors you need to consider when placing hives on farms include:

- Crops within 500 m of natural vegetation may already receive good numbers of wild bees.

- As the bees only forage 500 m, hives are best dispersed around the farm so the bees can reach all areas of the target crop.

- It is safe to keep hives close to places where people live and work.

- Stingless bees do not need access to water as honey bees do.

- Consider the predominant wind direction, especially during the flowering season. Bees tend to fly upwind from the hive to the crop, so concentrate the hives downwind of the target areas.

- Avoid drifting and fighting between hives by ensuring they are properly spaced.

- Place the hives in a suitable microclimate.

- Estimate an appropriate stocking rate.

- Minimise exposure to pesticides.

The last four factors are explained on the next two pages.

16.6.3 AVOID DRIFTING AND FIGHTING BETWEEN HIVES

When hives are placed in close proximity, foraging bees may **drift** (return to the wrong hive). The guard bees at the entrance recognise them as non-nestmates and repel or attack them. The defensive reactions can escalate into a large number of bees flying close together in a circular motion or hovering near the hive entrance. The defensive reaction may also end in fighting between bees and cause large losses of workers. (Read more about this in "Fighting swarms" in Chapter 13.)

When a colony is engaged in this defensive activity, normal foraging activity slows or stops and the colony becomes ineffective for pollination. Hives need to be spaced out to avoid drift. The locations should allow the foraging bees to easily recognise the hive of their origin and return to it. Hives can be marked with various shapes and colours to allow the bees to easily recognise them (Figure 16-14).

FIGURE 16-14 Stingless bee hives distributed on a farm. Note that they are widely spaced, placed unevenly and the roofs are painted different patterns and colours to help the foragers find their way back to the correct hive.

16.6.4 PLACE THEM IN THE BEST MICROCLIMATE

Hives need to be placed in situations that provide the best available climatic conditions. This may be a shaded or sunny position depending on the circumstances. Generally, hives prefer shady positions but, if flowering happens at a cooler time of year, then sunny positions will stimulate foraging.

On one strawberry farm in subtropical Australia, the farmer moves the hives into a sunny position for the full flowering period — an extended period of approximately six months in the coolest part of the year (Figure 16-15). Those positions are a poor choice in summer because they cause hive overheating in the full sun. Thus, for the non-flowering, hotter part of the year, hives are moved to another location and placed in the shade.

FIGURE 16-15 A stingless bee hive placed in a sunny position on a winter-flowering strawberry farm.

16.6.5 CHOOSE THE STOCKING RATE

The stocking rate (or number of hives required per unit of crop land) has rarely been calculated for stingless bee hives. Even for well-studied honey bees, recommended stocking rates are only rough estimates.

The number of hives required depends on many factors such as colony strength, weather, abundance of wild pollinating insects, duration of flowering, and preference by the particular type of bee for the particular crop. Even rough estimates are useful, so here I hazard a recommendation of 20 stingless bee hives per hectare for bee-dependent tree crops. This figure is a starting point that needs confirmation and would benefit from more research into the effects of critical variables.

For example, the population of wild insect pollinators will change this figure. If wild pollinator populations are high, then the introduction of more bees may have no beneficial effect. Chris Fuller, a crop consultant with a wealth of experience in using stingless bees on macadamia farms, recommends one hive per hundred trees in areas with moderate wild populations of bees. If the tree density is 400 per hectare, then his recommendation works out to be four hives per hectare.

Mark Greco and colleagues estimated one hive of *T. carbonaria* per 1,900 greenhouse capsicum plants for optimal pollination. In the field, this would translate to about 10 hives per hectare.

16.6.6 HOW TO MINIMISE THE EFFECTS OF PESTICIDES

Social bee colonies may be especially susceptible to pesticides because foraging bees can carry the poison back to the hive where it can kill nestmates. Pesticides include insecticides, fungicides and herbicides. Insecticides are the most toxic because they are designed to kill insects. Herbicides and fungicides may also harm bees.

Farmers can take actions and make decisions that minimise the impact of pesticides. Pest control options that are softer, or more environmentally friendly, such as biological and cultural control, should be considered first. Biological control options include releasing natural enemies of the pests. Pheromone traps are available and help control some pests. Cultural control includes such methods as the use of plant varieties that are resistant to pests. The number of sprays should be reduced by careful monitoring of crop pests and beneficial natural enemies of the pests, along with related factors such as weather conditions.

When insecticides are truly required, they should be applied at night or when wind speeds and insect activity are lowest. Insecticides should be selected that have minimal impact on bees. The farmer should do everything possible to decrease the drift of insecticides onto sites where wild or managed colonies are located.

Hives may need to be moved, covered or closed during periods of pesticide application (Figure 16-16). Closing the entrance will prevent the bees leaving and making contact with insecticides. Covering the hives will prevent insecticide coating the outside of the box, where the bees have a high probability of exposure. Moving the hives can be used to reduce exposure. Hives are best closed at night or during some other period of forager inactivity, then moved to a cool, dark place protected from insecticide drift. The hive closure should allow some air exchange. (Read more about closing hives in "Entrances and closures" in Chapter 9.)

The hives can be placed outside the crop rows to minimise exposure to insecticide sprays (Figure 16-17). Windbreaks or areas of natural vegetation may provide safe havens for hives.

FIGURE 16-16 These hives are placed in the tree rows and so need particularly careful protection from pesticides. They will be closed and covered with the adjacent drums for about 24 hours around the periods of pesticide application. IMAGE **BENJAMIN KALUZA**

FIGURE 16-17 Stingless bee hive on a strawberry farm. Note hive location in windbreak outside crop for protection from insecticides. Also see images on the right.

These hives are located in windbreaks outside the crop, to minimise exposure to insecticides.

Some stingless bee initiatives

In Australia, the booming interest in native stingless bees has fostered innovative initiatives and businesses run by both individuals and organisations.

Below, I present a few case studies of such projects. While this is by no means a complete list, it does provide examples of a good range of stingless bee initiatives.

Australian Native Bee Research Centre

This privately funded centre, run by Dr Anne Dollin and her husband, Les, has been vital to the development of the stingless bee "movement" in Australia.

With the help of Les, a skilled bushman, Anne has conducted painstaking research on many aspects of native bees, including revising the entire Australian stingless bee fauna. This fundamental taxonomic research, in collaboration with Professor SF Sakagami and Dr Claus Rasmussen, has underpinned much following stingless bee research and activity in Australia.

Not content with just doing research, Anne spreads information in a wonderful magazine called the *Aussie Bee Bulletin*, an important set of booklets on stingless bees, and Aussie Bee, the first website in Australia focusing on native bees. Aussie Bee remains one of the world's best websites on stingless bees. The quality of information in Anne's scientific and popular publications is first-class.

Ku-ring-gai Council

Ku-ring-gai Council in Sydney runs the WildThings program to promote urban wildlife. One of its most popular initiatives is a native bee program that gives residents a hive of *Tetragonula carbonaria*, with the caveat that the council has access to split the hive as required.

The program started in 2004 and, within 10 years, had placed 450 hives on properties throughout the area. Peter Clarke is the force behind WildThings, and has shown that it is possible to successfully breed colonies of stingless bees in Sydney at a rate that is similar to that enjoyed in warmer Queensland conditions. Peter has gained a lot of experience in keeping stingless bees at the lower limit of their climatic range.

In Peter's own words: "Splitting is hot, dirty work and incredibly rewarding. At the end of a day you feel like a midwife who has created something of inestimable value and, at the risk of making you (the reader) feel sick, we just love it. Connecting with Ku-ring-gai residents who love their bees, residents who are endlessly fascinated by their comings and goings, is very reaffirming. This program not only assists pollination at a time when *Apis* are under attack from serious pests but connects people to a little Aussie battler who, although small, is very important to the ecosystem. Anecdotal evidence abounds of crops of never seen before bounty, of people who have a much greater interest in insects around their property, of people who have reduced or eliminated chemical use around their home. We're connecting people to an unknown world, the world of beneficial insects."

Bee rescue services

In the mid-1990s, Rob Raabe, Cec Heather, Col Webb and Alan Waters started a service to rescue stingless bees from trees being cleared in the Ipswich area west of Brisbane. Peter Davenport implemented a similar service around the Gold Coast. They honed their skills by trial and error as few guidelines were then available. Over the years, they saved hundreds of hives, which they either left in the original logs or transferred into boxes. They also inspired others to save native bees from tree clearing. Now, operators such as Tobias Smith of Bee Aware Brisbane and Tony Harvey of Wide Bay Stingless Bees run rescue services using Facebook pages.

Stingless bee enterprises

More and more enterprising individuals are building businesses around stingless bees. Russell Zabel is a leading propagator and retailer of hives. Others such as Tony Goodrich breed hives for wholesale supply to retailers. Some retailers work online (e.g. City Chicks), while others have a shopfront at businesses like nurseries. Mark Grosskopf, Frank Adcock and Chris Fuller are building businesses around the supply of pollination services to farmers. Dan Heard and others specialise in building bee hives for sale. My own business, Sugarbag Bees, provides educational experiences such as workshops, seminars and school visits.

Resources

Following are some examples of the fantastic resources available for different forms of learning.

Websites

"BowerBird" is a socially interactive website for citizen science. Members can upload images and locations of wildlife. Other members will discuss their identification. Within BowerBird, the "Australian Bees" project has over 200 members and hundreds of sightings. Join BowerBird to see some spectacular photography, add your own photographs, and learn how to identify bees.

www.bowerbird.org.au

The Sugarbag website. The Sugarbag Team provide information, produce hives and related products, and offer educational workshops.

www.sugarbag.net

The Aussie Bee and Australian Native Bee Research Centre site is the premier website in Australia (and the world!?) on native bees. It provides a lot of free native bee information in addition to the booklet series mentioned above.

www.aussiebee.com.au

You can see native bees in a sanctuary close to Sydney at Honeycomb Valley Farm.

www.honeycombvalley.com.au/

Russell and Janine Zabel's website on Australian native bees provides a wealth of interesting information.

www.zabel.com.au

Mark and Kim Grosskopf were the first to provide a service renting stingless bee hives for pollination.

www.croppollination.com.au

Chris Fuller rents stingless bee hives for pollination.

www.nativebees.com.au

Steve Maginnity provides a number of educational resources on stingless bees.

www.stevesnativebees.com.au

Valley Bees is a great example of a strong and active community group.

www.mrccc.org.au/valley-bees/

Megan Halcroft's Bees Business.

www.beesbusiness.com.au

Facebook pages

- Australian Native Bee Network
- Sugarbag Bees
- Bee Aware Brisbane
- Wide Bay stingless bees
- Native stingless Bees for Bello
- aussiebeewebsite

YouTube videos

Design of an Australian native bee hive:
www.youtube.com/watch?v=hz3YsEDEQOM

Internal architecture of an Australian native bee hive:
www.youtube.com/watch?v=lsZQeerdjdI

Splitting an Australian native bee hive:
www.youtube.com/watch?v=Yhk7YvxRzT0

Extracting honey from an Australian native bee hive:
www.youtube.com/watch?v=WrkO6lhGSh4

Books

Australian Stingless Bees: A guide to Sugarbag Beekeeping, by John Klumpp. Published and distributed by Earthling Enterprises 2007:
www.earthling.com.au

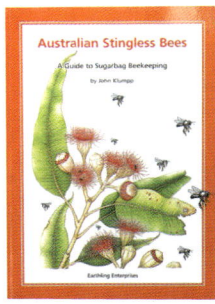

This 120-page book contains over 100 colour photos and presents a description and history of Australian stingless beekeeping in a wonderfully entertaining manner.

The magazines (*Aussie Bee Bulletin*) and booklets of the Australian Native Bee Research Centre are of an excellent standard and highly recommended. Visit:
www.aussiebee.com.au
for a list of information and products. Or write :
ANBRC, PO Box 74, North Richmond, NSW 2754

Buzzwords:

Important terms used in this book

Abdomen. Third and final part of the body of an insect.

Absconding. Departure of an entire colony of bees to a new nest site.

Advancing front. The part of the brood of stingless bees which advances because new cells are being built.

Age polyethism. The regular changing of labour roles by insect colony members as they age.

Aggregation. A group of individuals nesting in the same location but not cooperating to rear young.

Anther. The male part of the flower that produces and bears the pollen.

Anthophila. The scientific name for bees, also known as Apiformes.

Apine bees. Honey bees. Highly eusocial bees of the tribe Apini.

Apiculture. The keeping of honey bees.

Apiarist. A person who keep honey bees.

Batumen. The layer of material surrounding the nest of stingless bees comprised of propolis and debris.

Batumen plate. Thick batumen separating the nest of stingless bees from the larger cavity.

Bee bread. A mix of pollen and nectar deposited in a cell by a female solitary bee as food for her offspring.

Brood. A collective term for the immature stages (eggs, larvae and pupae) in a nest.

Brood cell. A cell constructed within a nest in which an individual bee is reared.

Brood chamber. A distinct area of the nest of stingless bees where the brood cells are concentrated.

Brood comb. Comb in which larvae are reared.

Brood food. Any food of larval bees.

Buzz pollination. Also known as sonication. The vibration of flowers by bees to dislodge pollen.

Callow. Also known as teneral, a young adult that has not attained mature coloration.

Castes. The functionally different groups of bees in a colony. In bees, the female castes are workers and queen. The male drones are also considered a caste.

Cells. Brood cells made of soft propolis in which a single bee is reared.

Cerumen. A material (mix of wax and plant resin) specific to stingless bee nests that is equivalent to propolis.

Cleptoparasite. A parasitic bee that lays eggs into the nest of other bees, usually of a different species.

Cluster. Refers to brood not in combs but irregularly arranged.

Cocoon. The protective covering made of silk spun by mature larvae of some bees.

Colony. The mature bees and larvae living and working in a nest.

Comb. A layer or layers of regularly arranged continuous cells.

Connectives. Elongate structures of propolis in the nest of stingless bees.

Corbicula (pl. corbiculae). The pollen basket on the hind legs of some bees, particularly honey bees, stingless bees and bumble bees.

Crop. An enlarged sac in the foregut of bees for transporting materials, particularly nectar.

Cross-pollination. Transfer of pollen from the flower of one plant to a different plant of the same species. In crops, cross-pollination has to be between plants of different varieties.

Diastase. An enzyme in honey responsible for the digestion of starch into sugars.

Drone. A male of a social bee.

Emergency queen cell. A cell for rearing a queen, constructed in a nest that has lost its queen.

Entrance. The external opening of a nest for the departure and arrival of bees. In stingless bees, it continues inside the nest cavity as an internal entrance tube and sometimes outside as the external entrance tube.

Eusocial. Living as a colony of multiple generations and castes.

Exoskeleton. An insect's tough outer shell that provides protection and support.

Family. A group of organisms that share similar characteristics. Animal families end in –idae and plant families end in –aceae. For example, the Colletidae is a family of bees that in Australia commonly forage on plants of the family Myrtaceae.

Feral honey bees. Wild bees that are not housed and managed in a hive.

Flower constancy. The tendency of a single bee to visit flowers of the same species while on a single foraging trip.

Forage. Verb, to collect nectar and pollen. Noun, natural sources from which bees collect food.

Forager. Also known as foraging bee or field bee. A bee that forages for nectar, pollen, resin or other materials and transports them back to the nest. Compare with House bee.

Genus (pl. genera). A group of closely related species. The genus name begins with a capital and is in italics. For example, *Austroplebeia* is a genus of stingless bees.

Haemolymph. The fluid that transports nutrients in the body of insects; it serves the same purpose as blood but is circulated in open spaces and is not restricted to vessels.

Haplodiploidy. The genetic system in which males have one copy of each chromosome (haploid) and are produced from unfertilised eggs, while females have two copies of each chromosome (diploid) and result from fertilised eggs.

Hibernation. A dormant stage that many animals enter to survive winter. In bees, it is often the pre-pupal or pupal stage. Stingless bees do not enter hibernation.

Highly eusocial. Living in a eusocial colony in which the castes differ in form.

Hive. An artificial home for a colony of bees.

Honey. A food produced by bees from nectar.

Honey bee. Correctly any bee of the genus *Apis*, but commonly only bees of the species *Apis mellifera*.

Honey pot. A chamber for storing honey in stingless bee nests.

Honey stomach. See Crop.

House bee. A worker bee that spends most of its time in the nest. Compare with Forager.

Hymenoptera. The order of insects that includes bees, ants and wasps.

Hypopharyngeal glands. Glands that open into the mouth of bees and secrete larval food and enzymes.

Involucrum. Thin layer or layers of soft propolis that surround the brood chamber of stingless bees.

Larva (pl. larvae). The young insect that hatches from the egg; it differs in form from the adult.

Life cycle. The sequence of life stages that an organisms undergoes; in bees it consists of egg, larva, pupa, adult.

Male congregation. Also known as lek or male swarm. A group of males that gathers to attract prospective partners for copulation.

Male swarm. See male congregation.

Mandibles. A pair of appendages near the insect's mouth, used for processing food, constructing a nest, and defending against predators or rivals.

Mandibular glands. Head glands that open into the mouth of bees and secrete food or pheromones.

Mass provisioning. Feeding larvae with sufficient food for the entire larval development period. Compare with Progressive provisioning.

Meliponary. A group of hives with bees, also known as stingless bee yard.

Meliponiculture. The keeping of stingless bees.

Meliponine bees. Stingless bees. Highly eusocial bees of the tribe Meliponini.

Meliponist. A person who keep stingless bees.

Melittology. The study of bees.

Metamorphosis. The change from larva to adult through the pupal stage.

Monogynous. The state in a social insect nest of having only one queen

Monolectic. Bees that collect pollen from only one species of plant. Compare with Polylectic.

Nectar. Sweet fluid secreted in the flowers of plants that attracts bees and is the material from which honey is made.

Nectary. A nectar-secreting gland.

Nest. A space or structure constructed by bees in which the adults live and the larvae are reared.

OATH with a view, Kin Kin, Queensland.
IMAGE **GLENBO CRAIG**

OATH. Original Australian Trigona Hive. A stingless bee hive design of standard dimensions that incorporates a method of colony propagation by horizontal splitting.

Ontogeny. The development of an individual from birth to death.

Ovule. The part of the flower, within the ovary, that can potentially develop into a seed.

Parthenogenesis. Development of an unfertilized egg into a new individual.

Perennial. (Of a plant or social insect colony) living for several or many years.

Pheromone. A substance used to communicate between individuals.

Phylogenetic tree. Also known as evolutionary tree, a branching diagram showing their phylogeny.

Phylogeny. The evolutionary relationships among species or groups of species, inferred from similarities and differences in their physical or genetic characteristics.

Physogastric. A word to describe a queen that has an abdomen enlarged with developing eggs.

Pistil. The female part of the flower bearing the stigma where the pollen grains are received and containing the ovules with the potential to grow into seeds.

Pollen. Microscopic grains that are the male gametes or sex cells of plants.

Pollen basket. See Corbicula.

Pollination. The transfer of pollen from the male to the female parts of flowers of the same species that results in the fertilisation of the ovules of the flower.

Pollinator. An animal agent such as a bee that transfers pollen for pollination.

Polylectic. A word to describe an animal that uses pollen from a wide range of plant species. Compare with Monolectic.

Pooter. Also known as aspirator, a device used in the collection of insects.

Pot. A structure used by stingless bees to store food, a pollen pot contains pollen, and a honey pot stores honey; together they can be called storage pots.

Pre-pupa. The last part of the last larval stage when the larva has defecated, ceased feeding and straightened.

Primitively eusocial. A eusocial state in which the castes are similar and adult food exchange is absent.

Progressive provisioning. Feeding of larvae at intervals during their growth. Compare with Mass provisioning.

Propolis. A mixture of wax and plant resins used by stingless bees and, to a minor extent, by honey bees, for nest construction.

Pupa (pl. pupae). The life stage in an insect in which its body form changes from larva to adult.

Pupation. The act of changing from lava to adult.

Queen. The female in a social insect colony that lays the eggs and does not forage.

Queen cell. The cell in which a new young queen is reared.

Queenless colony. A colony that has lost its queen.

Queenright colony. A colony with a mated, fertile, egg-laying queen.

Recruitment. The process by which foraging bees communicate their findings to other bees in the colony and stimulate them to forage on particular plants in the same area.

Reproductive swarm. The queen and workers that establish a new colony of bees.

Resin. A hydrocarbon secretion of many plants that hardens into transparent or opaque solids. It is distinct from other plant compounds such as sap, latex, or mucilage.

Royal jelly. A substance also known as bee milk, used to feed the queen and deposited in queen cells to feed the larvae of *Apis*.

Scopa (pl. scopae). The brush of hairs on the body of a bee in which pollen is carried from flowers to the nest.

Self-pollination. Transfer of pollen from the anther of a flower to the stigma of the same flower or another flower on the same plant.

Semi-social. A state of cooperation between colony members that is intermediate between solitary and eusocial.

Silk. Secretion of larvae that sets hard to form a cocoon.

Social. Referring to a group of bees that form a colony.

Solitary. Refers to bees that build their own nest and rear their own young without cooperating with others.

Sonication. See Buzz pollination.

Species (pl. species). A group of individuals that breed freely with each other.

Spermatheca. A reservoir in females in which sperm cells are stored after mating and from which they are released to fertilise the egg.

Split. To divide or propagate a colony to increase hive numbers.

Stigma. The sticky surface of the female part of the flower that is receptive to pollen.

Sting. An organ of defence in bees, consisting of a stinger shaft and venom sac.

Style. The part of the female flower in the form of a stalk that supports the stigma and provides the path for the pollen tube to reach the ovule.

Super. Section of hive for honey storage.

Supersedure. The replacement of an aging queen by a new one.

Swarm. A group of bees. For example, a reproductive swarm, absconding swarm, fighting swarm, male swarm.

Thorax. The second part of the insect body between the head and abdomen, where the wings and legs are attached.

Tibia. A segment of the leg. In some bees, the outer part of the hind tibia is modified to form the corbicula.

Tongue. An organ at the tip of the bee's proboscis, used to ingest liquids.

Trophallaxis. Interchange of material, particularly food, among members of a colony.

Virgin queen. Also known as gyne, a female bee that may become a queen.

Wasps. An insect of the order Hymenoptera and suborder Apocrita that is neither a bee nor an ant. Very common parasites or predators of other insect species.

Wax. Secretion from glands in the abdomen of certain bees used in building nests.

Worker. Member of a colony that works in the hive or forages and does not usually lay eggs.

Index